Contents

List of Tables

List of Figures

viii

xi

xiv

Chapter 1

Introduction

The purpose of this introduction is to provide a general perspective on control of wastewater nutrients like nitrogen and phosphorus. Nutrient control is not needed under all conditions, but it is considered mandatory for treating wastewaters of human origin that discharge directly or indirectly to lakes or reservoirs to prevent acceleration of eutrophication. It is an essential part of treating wastewaters used for recharge of groundwaters for public supplies. It also may be necessary when treating discharges to flowing waters, particularly in shallow streams where rooted aquatic plants can flourish, or where nitrification can tax oxygen resources.

To begin with, it is important to define two terms—eutrophication and eutrophic—as they apply to lakes, ponds, reservoirs, and other lentic or non-flowing waters.

Eutrophication is the term used to describe the natural process in which biological productivity increases with the age of a body of water. This is generally a result of capture by phytoplankton and other aquatic growths of plant nutrients contributed by inflowing waters and the new growth that results. These resultant organisms and plants eventually die and settle to the bottom, where they decompose to some degree. During decomposition, nutrients are released to the water above. These nutrients reach the upper waters or the euphotic zone in time, and this continual enrichment from external and recycled sources perpetually produces new growths that die and settle to the bottom. Residues from decomposition and silt carried in inflowing waters gradually fill the lake or reservoir.

This is known as the aging process that every body of water passes through on its way to extinction. During their lifetimes of tens or hundreds of thousands of years, bodies of water pass through three distinct phases. The first is the oligotrophic phase where biological productivity is low because of low nutrient loadings. As nutrient loadings increase, the mesotrophic phase of greater biological productivity develops. With greater and greater nutrient loadings from external sources and internal recycling, the third or eutrophic phase develops, with its attendant nuisance conditions caused by excessive biological activity. The purpose of nutrient control is to limit external sources as much as possible and thereby slow the eutrophication process and reduce its negative effects.

Eutrophic is the term applied to waters with a high degree of biological productivity. Several parameters are used to determine the trophic condition of a body of water: standing crop of phytoplankton, level of chlorophyll a, volume of algae, level of oxygen production, level of oxygen depletion, Secchi disk readings, or a combination of all these factors. In deep lakes that stratify during the summer, a common method of determination measures dissolved oxygen (DO) levels in the bottom layer, or hypolimnion. If the deep waters remain aerobic during summer stagnation, the lake is considered oligotrophic. If the DO becomes depleted, the waters are considered eutrophic, and the degree of eutrophy can be estimated by the time it takes for anoxic or anaerobic conditions to develop after the onset of stratification. Usefulness of this parameter requires some judgment based on temperature conditions and the relative volumes of the hypolimnion and the upper water layer, or epilimnion.

1

INTRODUCTION

Why have several countries embarked on programs of nutrient control that will cost billions of dollars? The justification for expenses of this magnitude is firm. Studies conducted throughout the world have shown that domestic and certain industrial wastewaters, and drainage from agricultural and urban areas have greatly accelerated the rate of eutrophication in receiving waters.

In considering nutrient removal or control, logical questions about the "what" and the "where" arise. It is appropriate to consider the "what" first.

Table 1.1 lists the primary nutrients for the production of algae. Actually, the prime source of carbon is carbon dioxide. As for all green plants, nitrogen is derived primarily from ammonia and nitrates. Phosphate ion is the sole source of phosphorus, and dissolved silicates are the sources of silicon. In addition, nitrogen-fixing algae are capable of using dissolved nitrogen gas in water, when ammonia and nitrates are in short supply. Silicon requirements range widely because most algae have little need for silicon, although diatoms use large amounts in forming their shells or encapsulations.

Because carbon dioxide is the most needed prime nutrient, it would seem that its control might serve as a means of limiting algal growths. But, such a concept fails to recognize the large stores of half-bound carbon dioxide that exist in most natural waters in the form of bicarbonates (alkalinity) that are available to algae. It also fails to recognize the fact that as carbon dioxide is abstracted from bicarbonates, pH value rises because of the shift to carbonate and hydroxyl ions, thus making carbon dioxide more readily absorbable from the atmosphere. Little, if any, credence is given to this concept today.

Nitrogen in the biomass of algae occurs in amounts ranging from 3 to 10%, largely in the form of proteins. Smaller amounts occur in green algae and larger amounts in blue-greens. However, because several blue-green algae can fix elemental nitrogen dissolved in water, fixed forms of nitrogen available from aqueous sources may not be true growth restraints. Numerous studies have shown that nitrogen can become limiting in the control of algal growths during the summer growing season at a limiting level of 0.05 mg/L of inorganic forms (NH_3-N + NO_3-N). On this basis, nitrogen removal from wastewaters may become necessary when receiving waters are insufficient in quantity to dilute inorganic forms to the limiting level.

Although phosphorus in algal cells occurs in small amounts ranging from 0.5 to 1.0% in the biomass, it has been shown to be a limiting factor in the growth of algae in many instances. A value of less than 0.005 mg/L or 5 ug/L in the ortho form is recognized as a lower growth limiting concentration.[1] Its removal from wastewaters is highly feasible because phosphates and most organic forms are readily removed by precipitation with use of alum, ferric salts, or lime. In addition, it can be removed by controlled biological treatment processes. Where land is available, it can be removed by application to cultivated, forested, and pasture areas.

In instances where nutrient control is deemed desirable, control of phosphorus is considered to be absolutely essential because, when nitrogen becomes limiting, any excess of phosphorus can support growth of nitrogen-fixing blue-green algae. In such cases, the nitrogen budget of a body of water will be increased, thereby materially off-setting any benefits from nitrogen removal.

Considering the above information,

TABLE 1.1. Primary nutrients for the production of algae.

Nutrient	Composition of algae biomass, %
Carbon	35–50
Nitrogen	3–10
Phosphorus	0.5–1.0
Silicon	0.1–14

many authorities believe the eutrophication problem can be controlled by rigid requirements on phosphorus removal alone. Others believe that a high degree of nitrification is also needed to eliminate detrimental side effects of ammonia nitrogen. The latter proposal meets with some objection on the part of people concerned with public water supplies because current water standards limit nitrate nitrogen to 10 mg/L.

As people concerned with eutrophication problems have become convinced that correction must depend on control of phosphorus and nitrogen, their attention, of necessity, has focused on the need for establishing a nutrient budget for each body of water. This involves an evaluation of the amounts contributed by all sources of water entering lakes and reservoirs—rivers, creeks, groundwater, atmosphere, rainfall, and wastewater effluents. Streams draining agricultural lands where inorganic fertilizers are used for fertility maintenance receive nutrients via underground and overland drainage carrying silt and nutrients off the land. In the same fashion, cropland receiving animal or green manure and grazing lands can also be sources of nitrogen and phosphorus. Agriculturally, it is impossible to attain a complete balance of fertilization with crop consumption of nutrients. Excess nitrogen must be applied to compensate for losses from denitrification and conversion to nitrates that are purged through the soil and lost in ground water. In the case of phosphorus, losses are primarily in silt carried away by wind and water erosion.

The establishment of a nutrient budget for a body of water is a time-consuming and expensive proposition. It usually requires a year-long study in foul and fair weather. Streams must be gauged and samples collected frequently during periods of storm runoff and nutrient contents of base flows established. Analytical procedures should distinguish between readily available and slowly available nutrients. They should also distinguish between insoluble and soluble materials because insolubles are apt to become part of bottom deposits.

In some instances, contributions from uncontrollable non-point sources may be so great that benefits from removal of nitrogen or phosphorus become insignificant. The purpose of establishing a nutrient budget is to prevent expenditure of funds for nutrient removal where little or no benefit can be expected. Careful evaluation of data is required because concentration is a factor that must be considered in conjunction with quantity.

This introduction would not be complete without some reference to the influence of nutrients on streams and the use of wastewater effluents for groundwater recharge. With respect to streams, ammonia nitrogen can cause serious oxygen sag problems when nitrification occurs rapidly. This problem is bound to occur in shallow streams, but becomes less and less of a problem as depth increases. A second problem in shallow streams is caused by development of rooted aquatic plants that restrict flow and tax oxygen resources at night. Both nitrogen and phosphorus contribute to this problem. In such instances, it is questionable whether phosphorus removal alone can correct the situation. However, because it can be accomplished in most cases with little capital investment, it should be tried before large sums are committed for nitrogen removal.

The degree of treatment of wastewaters needed to make them suitable for groundwater discharge varies considerably. Where large areas are available for surface percolation, effluent from biologically treated wastewaters is usually acceptable, as at Whittier Narrows, Calif. In situations where percolation areas are not available, it becomes necessary to resort to injection procedures. Under such conditions, it is normally necessary to remove nitrogen and phosphorus and a large part of the total solids.

In areas where suitable land is avail-

INTRODUCTION

able, essentially complete or partial removal of both nitrogen and phosphorus can be accomplished. In situations where wastewaters are used for irrigation purposes, there is normally no runoff and all of the phosphorus is removed as the water percolates through the soil. In the case of nitrogen, however, any nitrogen in excess of plant requirements eventually will appear as nitrates and be washed through the soil to appear in groundwater or drainage. Partial removal is accomplished by overland flow schemes with continual runoff carrying any nutrients not removed by soil contact.

REFERENCES

1. "Water Quality Criteria." Federal Water Pollution Control Admin., Supt. of Documents, U.S. Gov. Printing Office, Washington, D.C. (April 1968).

4

Chapter 2
Source and Quantity of Nutrients

PHOSPHORUS

Phosphorus is an essential nutrient for the growth of plants and microorganisms. However, excess phosphorus, and nitrogen, in treated wastewater has been shown to be associated with undesirable algae and plant growth in receiving waters.[1-3] The effects of excess phosphorus on receiving waters are detailed in Chapter 4.

The major sources of phosphorus in domestic wastewater are human excrement, synthetic laundry detergents, and water treatment chemicals.[4] Phosphorus contributed by detergents is reduced significantly in communities that enforce "phosphate bans." In areas with such "phosphate bans," treatment plant effluents have 50% less total phosphorus than comparable treatment systems operating in areas without restrictions on laundry detergents. Use of sodium metaphosphate or similar compounds (for corrosion control) in the potable water supply system also will increase the level of phosphorus in a wastewater.

Typical phosphorus concentrations for fresh domestic wastewater are listed in Table 2.1. The phosphorus is listed as total phosphorus, organically bound phosphorus, and inorganic phosphorus, all expressed as P. The inorganic phosphorus includes simple orthophosphates and polyphosphates. Phosphorus quantities are expressed as mass phosphorus per unit volume, (mg/L and lb/mil gal). The inorganic forms of phosphorus comprise approximately 70% (4 to 15 mg/L) of the 6 to 20 mg/L total phosphorus

present in domestic wastewater. The remaining 2 to 5 mg/L is organically bound phosphorus. Table 2.1 includes a per capita generation rate of phosphorus. The generation rate of total phosphorus is 0.8 to 1.8 kg/cap·a (1.8 to 4.0 lb/cap/yr), with 0.5 to 1.2 kg/cap·a (1.1 to 2.7 lb/cap/yr) being inorganic phosphorus.

Industry can either add, or in the case of a few phosphorus-deficient discharges, dilute the total phosphorus in wastewater. Industrial wastes typically high in phosphorus include those generated from fertilizer production, feedlots, meat processing and packing, milk processing, commercial laundries, and some food processing wastes. Certain pulp and paper manufacturing processes discharge a phosphorus-deficient waste.

The soluble orthophosphates are the simplest forms of phosphorus and are the end products of the breakdown of inorganic polyphosphates and organically bound phosphorus. Soluble orthophosphates typically comprise 15 to 35% of the total phosphorus in fresh domestic wastewater. The ratio of orthophosphates to polyphosphates and organically bound phosphorus is dependent both on characteristics of the wastes discharged to the wastewater stream, and the degree to which the complex forms have broken down.

Treatment of wastewater increases the soluble orthophosphate level to an estimated 50 to 90% of the total phosphorus. The soluble orthophosphate form of phosphorus is both the easiest to pre-

TABLE 2.1. Phosphorus in domestic wastewater.

| | Concentration as P[a] | | Phosphorus generation |
	mg/L	lb/mil. gal	kg/cap · yr (lb/cap/yr)
Total	6–20	50–165	0.8–1.8 (1.8–4.0)
Organic	2–5	17–42	0.3–0.6 (0.7–1.3)
Inorganic	4–15	33–125	0.5–1.2 (1.1–2.7)

[a] Typical concentrations in fresh domestic wastewater.

cipitate and the form most available for assimilation by algae and plants. The effluent of biological treatment systems will contain some organically bound phosphorus in cell material.

Phosphorus removal methods are detailed in Chapter 7 and in other texts.[6,7] Conventional processes usually involve precipitation with alum or ferric chloride. Percent removals (of influent concentration) of total phosphorus obtainable with primary and secondary treatment, both with and without chemical (alum or ferric chloride) addition, are presented in Table 2.2. Primary and secondary treatment without chemical precipitation removes little of the phosphorus: 5 to 10% with primary, and 10 to 20% with secondary treatment. With addition of alum or ferric chloride, the removals increase to 70 to 90% with primary, and 80 to 95% with secondary treatment levels. Effluent concentrations of 1 mg/L total phosphorus are attainable with secondary treatment. Table 2.2 also includes a percent removal (80%) for primary treat-ment with lime addition. Higher levels of phosphorus removal can be obtained with advanced wastewater treatment (discussed in Chapters 7 and 8).

NITROGEN

Nitrogen is another nutrient essential for the growth of plants and microorganisms. However, excess nitrogen in wastewater treatment plant effluents, like excess phosphorus, has been shown to be associated with undesirable algae and plant growth in receiving waters. And, the ammonia form of nitrogen is toxic to fish.[1-3,5] The effects of excess nitrogen on receiving waters are detailed in Chapter 4.

Nitrogen enters the domestic wastewater stream primarily as urea and combined in feces and other organic material. Urea has often been considered an organic nitrogenous compound, but it is actually a derivative of carbon dioxide (as in limestone) and should be classified as inorganic. The urea is rap-

TABLE 2.2. Typical removal efficiencies, total phosphorus.

| | Percent removal of influent concentration | | |
Treatment level	Without chemical[a] addition	With chemical[a] addition	Lime addition
Primary	5–10	70–90	80
Secondary	10–20	80–95	

[a] Typically alum or iron (ferric chloride).

TABLE 2.3. Nitrogen in domestic wastewater.

| | Concentration (as N)[a] | | Generation |
	mg/L	lb/mil. gal	kg/cap·yr (lb/cap/yr)
Total	20–85	167–709	3.4–5.0 (7.5–11)
Organic	8–35	67–292	0.7–1.0 (1.5–2.2)
Free ammonia/ammonium	12–50	100–417	2.7–4.0 (4.8–8.8)

[a] Typical concentrations in fresh domestic wastewater.

idly hydrolyzed by enzymes to ammonia (or the ammonium ion, depending on pH) and carbon dioxide.

Typical nitrogen concentrations for fresh domestic wastewater are listed in Table 2.3. The nitrogen is listed as total nitrogen, organic nitrogen, and ammonia/ammonium nitrogen. Two other forms of nitrogen are present in domestic wastewater: nitrite and nitrate nitrogen. However, in fresh domestic wastewater the concentrations of these (especially nitrite) are minimal. An exception to this can be found in communities with high nitrate levels in their drinking water. In Table 2.1, the nitrogen quantities are expressed as mass nitrogen per unit volume (mg/L and lb/mil gal). The ammonia form of nitrogen comprises approximately 60% (12 to 50 mg/L) of the 20 to 85 mg/L of total nitrogen present in fresh domestic wastewater. The remaining 8 to 35 mg/L is organically bound nitrogen. Table 2.3 includes a per capita generation rate of nitrogen. Total nitrogen generation rate is 3.4 to 5.0 kg/cap·a (7.5 to 11.0 lb/cap/yr), with 2.7 to 4.0 kg/cap·a (4.8 to 8.8 lb/cap/yr) being ammonia.

Industrial discharges can add significant quantities of nitrogen to the wastewater stream. Industrial wastes typically high in nitrogen include those generated from feedlots, fertilizer production, meat processing, milk processing, petroleum refineries, coking facilities, certain synthetic fiber plants, and industries that clean with ammonia-containing compounds.

The form of nitrogen is an indicator of the age and condition of the wastewater. Fresh wastewater, as noted, contains primarily ammonia (from the rapidly hydrolyzed urea) and organic compounds of nitrogen (proteins, peptides, amino acids, creatine, uric acid, and others). The organic compounds are slowly decomposed by bacteria to ammonia, carbon dioxide, and water. Under aerobic conditions, the ammonia form of nitrogen is oxidized to nitrite (NO_2) by the *Nitrosomonas* genera of bacteria. The nitrite nitrogen is an unstable compound that is rapidly further oxidized by *Nitrobacter* bacteria to nitrate nitrogen (NO_3). The oxidation of ammonia to nitrite to nitrate is called nitrification (detailed in Chapter 5) and is the basis of ammonia removal in wastewater treatment. Nitrification is an oxygen intensive process

TABLE 2.4. Typical removal efficiencies, total nitrogen.

Treatment level	Percent removal of influent concentration
Primary	5–10
Secondary	10–20
Secondary with nitrification	20–30

theoretically requiring 4.6 mg/L of oxygen to nitrify 1 mg/L of ammonia.

Nitrogen removal processes are detailed in Chapter 6. As noted, conventional processes usually involve a planned nitrification step. Percent removals (of influent concentration) of total nitrogen without advanced wastewater treatment are listed in Table 2.4. Typical removals are listed for primary and secondary treatment levels and secondary treatment with nitrification. Nitrogen removals without advanced treatment are low: 10 to 20% for secondary treatment and 5 to 10% for primary treatment levels. Secondary treatment with nitrification can obtain removals of up to 30%. Effluent limitations for most receiving waters will require advanced nitrogen removal processes. Advanced nitrogen removal is discussed in Chapters 6 and 8 of this manual.

REFERENCES

1. Gakstatter, J. H., *et al.*, "A Survey of Phosphorus and Nitrogen Levels in Treated Municipal Wastewater." *J. Water Pollut. Control Fed.*, **50**, 718 (1978).
2. "Water Quality Improvement by Physical and Chemical Processes." E. F. Gloyna, W. W. Eckenfelder, Jr., (Eds.), Water Resources Symposium No. 3, Center for Research in Water Resources, University of Texas Press, Austin (1970).
3. Kuznetsov, S. I., "The Microflora of Lakes and its Geochemical Activity." University of Texas Press, Austin (1970).
4. "Process Design Manual for Phosphorus Removal." U.S. EPA, Washington, D.C. (April 1976).
5. "Process Design Manual for Nitrogen Control." U.S. EPA, Washington, D.C. (Oct. 1975).
6. "Wastewater Engineering: Collection, Treatment, Disposal." Metcalf & Eddy, McGraw Hill Book Co., New York, N.Y. (1972).
7. "Wastewater Treatment Plan Design." Manual of Practice No. 8, Joint Publication of ASCE and WPCF, Water Pollut. Control Fed., Washington, D.C. (1977).

Chapter 3

Analytical Methods

Instruments and instrumental techniques are available for measuring and detecting many of the nutrients present in wastewaters or receiving water bodies. Some chemical analysis methods have been automated: alkalinity, ammonia, calcium, chemical oxygen demand (COD), chloride, fluoride, hardness, nitrite, nitrate, pH, phosphate, silica, sulfate, and various metals by colorimetry. Some of these constituents can be determined automatically, either by measurement of the color developed in a treated sample, or by colorimetric determination of a titration end point.

The increasing availability and variety of automated analytical instrumentation for nutrient determination make giving detailed descriptions and operating instructions for all of them impossible. But, it is important for the analyst to recognize that the chemical principles on which automated methods are based are the same as, or comparable to, the chemical principles governing manual methods. In some instances, a method differing from the manual method is chosen for automation because of simplicity or stability. And, sometimes a loss of sensitivity results. In no case can an automated instrument improve a method that is analytically unsuitable for the measurement required. However, it is possible to use, in automated versions, methods that would be difficult to perform manually. For example, where the time period between reagent additions and colorimetric measurement must be precisely the same from sample to sample, an automated method may be better than a manual technique.

PHOSPHORUS

Phosphorus analyses embody two general procedural steps: conversion of the phosphorus form of interest to dissolved orthophosphate, and colorimetric determination of dissolved orthophosphate. The separation of phosphorus into its various forms is defined analytically, but the analytical differentiations have been selected so that they may be used for interpretive purposes.

Filtration through a 0.45-μm membrane filter separates "filterable" from "nonfilterable" forms of phosphorus. Note that no claim is made that filtration through 0.45-μm filters is a true separation of suspended and dissolved forms of phosphorus. It is merely a convenient and replicable analytical technique designed to make a gross separation. This is reflected in the use of the term "filterable," rather than dissolved, to describe the phosphorus forms determined in the filtrate that passes the 0.45-μm filter.

Membrane filtration is selected over depth filtration because of the greater likelihood of obtaining a consistent separation of particle sizes. Prefiltration through a glass fiber filter may be used to increase the filtration rate as well as to improve the consistency of size separation.

Phosphates that respond to colorimetric tests without preliminary hydrolysis or oxidative digestion of the sample are termed "reactive phosphorus." While reactive phosphorus is largely a measure of orthophosphate, a small fraction of any condensed phosphate present is usually unavoidably hydrolyzed in the procedure. Reactive phosphorus occurs in both filterable and nonfilterable forms.

Acid hydrolysis at boiling-water temperature converts filterable and particulate condensed phosphates to filterable orthophosphate. The hydrolysis unavoidably releases some phosphate from organic compounds, but this may be reduced to a minimum by judicious selection of acid strength and hydrolysis time and temperature. The term "acid-hydrolyzable phosphorus" is preferred over "condensed phosphate" for this fraction.

The phosphate fractions that are converted to orthophosphate only by oxidative destruction of organic matter present are considered "organic" or "organically bound" phosphorus. The severity of the oxidation required for this conversion depends on the form—and to some extent on the amount—of the organic phosphorus present. Like reactive phosphorus and acid-hydrolyzable phosphorus, organic phosphorus occurs both in the filterable and nonfilterable fractions. With minor variations, the filterable and nonfilterable fractions of a sample correspond to dissolved and particulate phosphates, respectively.

The total phosphorus, as well as the filterable and nonfilterable phosphorus fractions, each may be divided analytically into the three chemical types that have been described; reactive, acid-hydrolyzable, and organic phosphorus.

Figure 3.1 shows the steps for analysis of individual phosphorus fractions. As indicated, determinations usually are conducted only on the unfiltered and filtered samples. Non-filterable fractions generally are determined by difference.

SELECTION OF METHOD

Digestion methods. Because phosphorus may occur in combination with organic matter, a digestion method to determine total phosphorus must be able to oxidize organic matter effectively to release phosphorus as orthophosphate. Three digestion methods are given. The perchloric acid method, the most meticulous and time-consuming method, is recommended only for particularly difficult samples such as sediments. The nitric acid-sulfuric acid method is recommended for most samples. By far the simplest method is the persulfate oxidation technique. It is recommended that this method be checked against one or more of the more drastic digestion techniques and be adopted if identical recoveries are obtained.

Colorimetric methods. Three methods of orthophosphate determination are described in the Standard Methods.[1] Selection depends largely on the concentration range of orthophosphate. The vanadomolybdic acid method is most useful for routine analyses in the range of 1 to 20 mg P/L. The stannous chloride method or the ascorbic acid method is more suited for the range of 0.01 to 6 mg P/L. An extraction step is recommended for the lower levels of this range and when interferences must be overcome.

Automated methods. Two automated systems that employ the same principle, but differ in detail and rate of analysis, are in common use. They consist of the components shown in Figures 3.2 and 3.3.

NITROGEN

In waters and wastewaters, the forms of nitrogen of greatest interest are, in

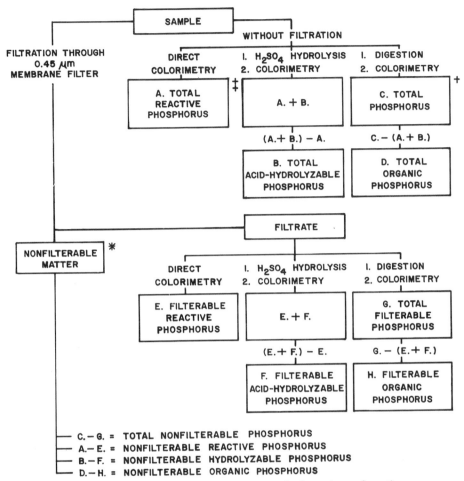

FIGURE 3.1.   Steps for analysis of phosphate fractions.

order of decreasing oxidation state: nitrate, nitrite, ammonia, and organic nitrogen. All these forms of nitrogen, as well as nitrogen gas (N_2), are biochemically interconvertible and are components of the nitrogen cycle. They are of interest for many reasons. Organic nitrogen is defined functionally as organically bound nitrogen in the trinegative oxidation state. It does not include all organic nitrogen compounds. Analytically, organic nitrogen and ammonia can be determined together and have been referred to as "kjeldahl nitrogen," a term that reflects the technique used in their determinations. Organic nitrogen includes such natural materials as proteins and peptides, nucleic acids and urea, and numerous synthetic organic materials. Typical organic nitrogen concentrations vary from a few hundred micrograms per litre in some lakes to more than 20 mg/L in raw wastewater.

Total oxidized nitrogen is the sum of nitrate and nitrite nitrogen. Nitrate generally occurs in trace quantities in surface water, but may attain high levels in some groundwater. Nitrate is found only in small amounts in fresh domestic wastewater, but in the effluent of nitrifying biological treatment plants, nitrate may be found in concentrations of up to 30 mg/L as nitrogen.

Nitrite is an intermediate oxidation

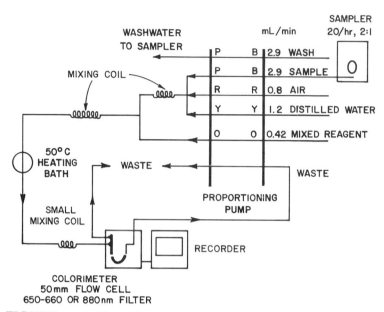

FIGURE 3.2. Phosphate manifold for Automated System I.

state of nitrogen, both in the oxidation of ammonia to nitrate and in the reduction of nitrate. Such oxidation and reduction may occur in wastewater treatment plants, water distribution systems, and natural waters.

Ammonia is naturally present in surface and wastewaters. Its concentration generally is low in groundwaters because it adsorbs to soil particles and clays and is not readily leached from soils. It is produced largely by the deamination of organic nitrogen-containing compounds and by the hydrolysis of urea.

In the chlorination of wastewater effluents containing ammonia, virtually no free residual chlorine is obtained until the ammonia has been chlorinated. Rather, the ammonia reacts partially with chlorine to form mono- and dichloramines. Ammonia concentrations encountered in water vary from less than 10 μg/L as ammonia nitrogen in some natural surface and groundwaters to more than 30 mg/L in some wastewaters.

Nitrogen (ammonia). The two major factors that influence selection of the method to determine ammonia are concentration and presence of interferences. In general, direct manual determination of low concentrations of ammonia is confined to drinking waters, clean surface water, and good-quality nitrified wastewater effluent. In other instances, and where interferences are present and greater precision is necessary, a preliminary distillation step is required. For high ammonia concentrations a distillation and titration technique is preferred.

Two manual colorimetric techniques—the nesslerization and phenate methods—and one titration method are suggested in "Standard Methods."[1] An ammonia-selective electrode method, which may be used either with or without prior sample distillation, and an automated version of the phenate method also are included. While the stated maximum concentration ranges for the manual methods are not rigorous limits, titration is preferred at concentrations higher

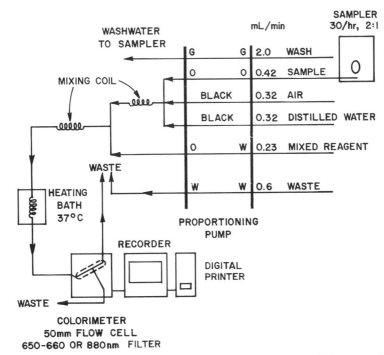

FIGURE 3.3. **Phosphate manifold for Automated System II.**

than the stated maximum levels for the photometric procedure.

Nitrogen (nitrate). Determination of nitrate (NO_3-) is difficult because of the relatively complex procedures required; the high probability that interfering constituents will be present; and, the limited concentration ranges of the various techniques. Two screening techniques for determining approximate NO_3- concentration are suggested in "Standard Methods."[1] A choice of techniques, depending on concentration range and presence of interferences, is also presented.

Screening methods include: an ultraviolet (UV) technique that measures the absorbance of NO_3- at 220 nm and is suitable for uncontaminated (low in organic matter) waters; and an NO_3- electrode method that may be used in either unpolluted water or wastewater.

Nitrogen (nitrite). Nitrite (NO_2-) is determined through formation of a reddish purple azo dye produced at pH 2.0 to 2.5 by coupling diazotized sulfanilic acid with N-(1-naphthyl)-ethylenediamine dihydrochloride (NED dihydrochloride). The method is suitable for determination of NO_2- down to 1 $\mu g/L$ NO_2-N. Photometric measurements can be made in the range of 5 to 50 $\mu g/L$ N if a 5-cm light path and a green color filter are used. The color system obeys Beer's law up to 180 $\mu g/L$ N with a 1-cm light path at 543 nm. High NO_2- concentrations can be determined by diluting a sample to 50 mL in the nessler tube used to conduct the reaction.

Nitrogen (organic). The major factor that influences the selection of a macro- or semi-micro-Kjeldahl method to determine organic nitrogen is the concentration of organic nitrogen. The

13

macro-Kjeldahl method is applicable for samples containing low concentrations of organic nitrogen and requires a relatively larger sample volume than the semi-micro-Kjeldahl method, which is applicable to samples containing high concentrations of organic nitrogen. In the tentative semi-micro-Kjeldahl method, a sample volume should be chosen that contains organic plus ammonia nitrogen (Kjeldahl nitrogen) in the range of 0.2 to 2 mg.

REFERENCES

1. "Standard Methods for the Examination of Water and Wastewater." 15th Ed., Am. Public Health Assoc., Washington, D.C. (1981).

Chapter 4

Role and Impact of Nutrients

AQUATIC ENVIRONMENT

The modern day concept of the role of nutrients in the aquatic environment has been a subject of considerable controversy for decades. Why the translation of well accepted principles of fertilization from land to water has been so slow is difficult to understand, especially considering the important role photosynthesis plays in both instances.

One undebatable fact concerning phytoplankton and aquatic plants was established in 1922 by Birge and Juday.[1] Using the classic method of Justus von Liebig, the famous agricultural chemist, they determined the elemental composition of such growths. A sample of their data is presented in Table 4.1. Of particular interest from a nutritional viewpoint is the high nitrogen content (protein) of the blue-green algae and the nitrogen to phosphorus ratios that do not exceed 15 to 1. These data illustrate the need for carbon (carbon dioxide) and the common fertilizing elements, nitrogen and phosphorus, as nutrients for aquatic growths.

Further evidence to support the conclusion that nitrogen and phosphorus are prime nutrients comes from numerous experiments where these elements were added to lakes and ponds to improve biological production as a means of increasing fish production. Also, all liquid media used in culturing algae include both nitrogen and phosphorus in their formulations. Still further proof, in an inverse way, comes from situations where diversions of wastewater effluents from critical aquatic areas, or actual removal of nutrients prior to release to such waters, has led to reduced biological productivity and relief from undesirable conditions.

Other major nutrients are carbon (carbon dioxide), sulfur (sulfates), calcium, magnesium, potassium, sodium, and silicon. Carbon dioxide is considered uncontrollable because of the large stores available to plant life as bicarbonates in most natural waters and in the atmosphere. Of the other elements, plentiful supplies exist in most waters, and no control could be expected by their removal from wastewaters.

Iron, manganese, zinc, molybdenum, and other materials are considered essential trace elements. These are required in such small amounts that little or no consideration is given to their control, in relation to biological productivity.

FLOWING WATER BODIES

Ammonia nitrogen. Ammonia nitrogen entering receiving waters behaves in

TABLE 4.1.   Analyses of some typical green algae, blue-green algae, and rooted aquatics.

Plant	Percent, dry mass			
	Carbon	Nitrogen	Phosphorus	N/P
Blue-green algae				
Anabaena	49.7	9.43	0.77	12/1
Aphanizomenon	47.7	8.57	1.17	7/1
Microcystis	46.5	8.08	0.68	12/1
Green algae				
Cladophora	35.3	2.30	0.56	4/1
Pithophora	35.4	2.57	0.30	8/1
Spirogyra	42.4	3.01	0.20	15/1
Rooted aquatics				
Elodea	—	2.10	0.14	15/1
Lobelia	—	1.89	0.16	12/1
Potomogeton	—	3.19	0.30	11/1

one of three ways. It may serve as a prime nutrient to support phytoplankton and rooted aquatic growths. Some will be lost to the atmosphere as gaseous NH_3 when its partial pressure in the water is greater than that of ammonia in the atmosphere. The degree of loss is a function of pH, which controls the concentration of ammonia (ammonium ion) by its effect on the equilibrium:

$$NH_4^+ \rightleftharpoons NH_3 + H^+ \qquad (1)$$

The significance of pH in the range of interest in natural waters is shown in Figure 4.1. At pH levels of 9 or above, significant losses can occur, but very little loss will occur at pH values below 7.

At this point it would be well to point out the need to ensure control of molecular ammonia concentrations to prevent fish toxicity.

Molecular or free ammonia in concentrations above 0.2 mg/L can cause fatalities in several species of fish. Applying the usual safety factor of 10, a National Academy of Sciences/National Academy of Engineering Committee has recommended that no more than 0.02 mg/L of free ammonia be permitted in receiving waters.[2]

The most notorious effect of ammonia nitrogen is related to its effect on the oxygen resources of a stream of river,

commonly referred to as nitrogenous oxygen demand (NOD). It is well known that ammonia nitrogen can be readily converted to nitrite and then nitrate, provided the necessary nitrifying bacteria are present.

$$2\,NH_3 + 3\,O_2 \xrightarrow{\text{Bact.}} 2\,NO_2^- + 2\,H_2O + 2\,H^+ \quad (2)$$

$$2\,NO_2^- + O_2 \xrightarrow{\text{Bact.}} 2\,NO_3^- \qquad (3)$$

The manner in which these reactions occur is displayed in Equations 2 and 3 and the rate of the reaction is in direct proportion to the population of nitrifying bacteria. Thus, it is possible that the reaction may be so slow that the rate of stream reaeration offsets the demand for oxygen and no noticeable oxygen sag will occur. On the other hand, where significant populations do occur, the demand for oxygen may be far in excess of natural reaeration capacity and a distinct oxygen sag will develop, as demonstrated in Figure 4.2.

The reason for the difference in the behavior of ammonia in rivers has to be related to difference in nitrifying populations. Usually, extensive and rapid nitrification has been noted in shallow streams and very low rates have been noted in deep ones.

At the present time, the accepted hy-

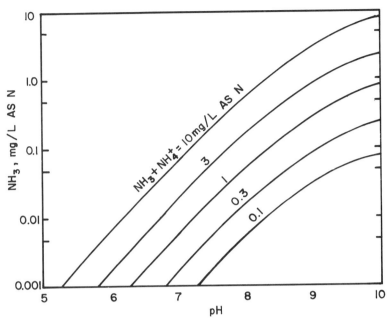

FIGURE 4.1. The effect of pH and ammonia nitrogen concentration (NH_3 + NH_4^+) on the concentration of free or molecular ammonia in water.

pothesis to explain this anomoly is as follows: In natural waters, flowing or otherwise, the population of suspended nitrifying bacteria is very small. Because nitrifying bacteria are slow growing organisms, it is impossible for them to reproduce fast enough to create condi-

tions as shown in Figure 4.2 within any short distance downstream of a point source. Because the normal habitats for nitrifying bacteria are attached growths, they occur in great populations in slimes on rocks and other surfaces. For this reason, the nitrification rate in a stream

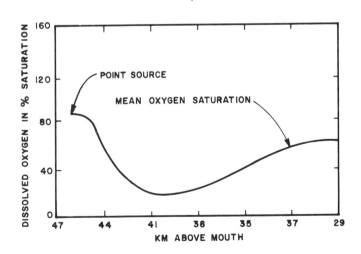

FIGURE 4.2. Dissolved oxygen sag caused by nitrification. Clinton River, Mich. Aug. 23–25, 1960.

is a function of the ratio of water volume to surface contact area. This explains why nitrification is rapid in shallow streams and very slow in deep ones. In effect, nitrification rate, like reaeration, is inversely related to depth. This hypothesis finds strong confirmation in a study conducted on the Willamette River, Oreg., by the U.S. Geological Survey.[3] Problems involving NOD can be solved by conversion of all ammonia to nitrates in the treatment plant.

Nitrogen and phosphorus. The fertilizing effects of nitrogen and phosphorus in receiving waters vary markedly in different depths. In deep rivers, there is stimulation of phytoplankton. Rooted aquatics are stimulated in shallow waters, along shorelines, and in backwaters.

In general, the effects are not serious in deep waters because of widely fluctuating water levels and flushing actions that occur at high river stages. But, the situation is different in shallow streams where rooted aquatics gain enough sunlight to thrive. Stimulation of growths can be so severe that natural flow may be impeded and backflooding may occur. If this happens in a stream draining from a lake, water levels in the lake can become uncontrollable. Dredging to remove the aquatic growths may become necessary. In some instances, severe diurnal fluctuations in dissolved oxygen can occur in stream waters passing through weed beds, as shown in Figure 4.3.

The data in Figure 4.3 were obtained at three stations downstream of a wastewater treatment plant. Station "A" was first in line of flow and received the full impact of residual biochemical oxygen demand (BOD) plus ammonia and phosphorus. Stations "B" and "C" were further downstream and were spared the full load of BOD that Station "A" received. During the day, submerged, in-stream growths produced oxygen in excess of the demands imposed by the BOD and NOD of the treatment plant effluent, particularly at "B" and "C". At night, the demands imposed by the effluent were increased by respiration of the plants and dissolved oxygen levels fell so low that they would not support fish life. This caused a catastrophe at a downstream fish hatchery that used diverted stream waters.

STATIONARY WATER BODIES

The most damaging and obvious effects of nutrients in lakes, ponds, and reservoirs are stimulation of algal growths that usually occur in pulses commonly referred to as blooms. Of particular concern are algae such as *Spirogyra, Anabaena, Aphanizomenon, Gleotrichia,* and others that tend to float to the surface at certain stages in their life cycles and are wafted about by gentle breezes to form accumulations in coves and along shore lines. Such aggregations can interfere with normal recreational activities such as swimming and cause serious nuisances because of death, subsequent odorous bacterial decomposition, and unsightliness. Odor problems can be particularly offensive when high-protein, blue-green algae, such as *Anabaena, Aphanizomenon,* and *Gleotrichia,* are involved.

Algal bloom nuisances have sparked legal controversies in the past. Perhaps the classic case was one in Madison, Wis., involving lake property owners, the state legislature, and the courts. Deliberations went on for 18 years before the Metropolitan Sewerage District finally was forced to divert its wastewater treatment plant effluent away from downstream lakes.

The algal bloom phenomenon is a simple biological process. Death is the ultimate fate of photosynthetic plants such as algae and rooted aquatics that thrive on plant nutrients. At the end of their lives, they settle in with previous crops in bottom sediments, where they then undergo bacterial decomposition. In decomposing, under aerobic conditions, they consume oxygen, which

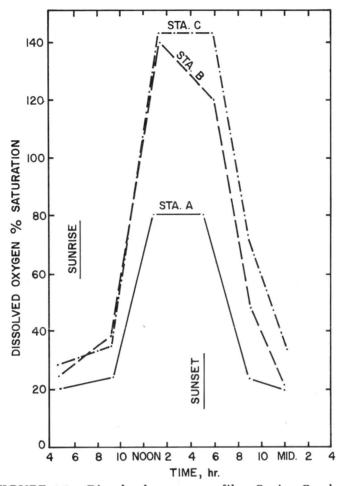

FIGURE 4.3. Dissolved oxygen profiles. Spring Creek, Pa.

eventually becomes depleted. This condition is quite normal in lakes that are deep enough to stratify. When oxygen is reduced below certain levels, the effects on fish life and other oxygen-dependent organisms are disastrous. Nitrates in the water may temporarily prevent the formation of anaerobic conditions, but their subsequent depletion will allow anaerobic conditions to develop. Release of hydrogen sufide can also occur from reduction of sulfates.

Bacterial decomposition under both aerobic and anaerobic conditions then leads to a phase that perpetuates the blooming process. The decaying bacteria release nitrogen as ammonia and phosphorus as phosphates. Small amounts also are released in soluble organic compounds and a fraction of the nutrients always remains as undigestible residues in bottom deposits. But the predominant release is of prime nutrients, which are then recycled to waters above.

In bodies of water that do not stratify, all water is considered in complete circulation as a result of wind action. Because of this mixing, nutrients released from bottom deposits are quickly carried back to waters in the euphotic zone where they become free to support new growths of photosynthetic organisms. In deep lakes that do stratify during the summer, the nutrients released are cir-

19

culated and stored within the hypolimnion (Figure 4.4). Small amounts may diffuse through the thermocline to the epilimnion and reach the euphotic zone. But, the major part of the nutrients released remains effectively sealed off until the fall overturn when the hypolimnion is disrupted by falling water temperatures and circulation patterns caused by strong winds. The nutrients then become distributed throughout the entire body of water. This seasonal enrichment of waters in the euphotic zone often results in fall blooms of algae.

The most damaging potential effect of nutrient releases like this will occur in lakes, ponds, and reservoirs with long retention periods. Only ponds without outlets can be considered strictly stationary, but all lakes and reservoirs have finite flows over time that range from a few days to many years. In general, at least an average 20-day detention time is required for released nutrients to develop serious algal growths. At shorter retention times, bodies of water assume characteristics similar to rivers.

Role as sedimentation basins. A common characteristic of all stationary bodies of water is that they serve as giant sedimentation basins that remove and store suspended matter from inflowing waters. This matter adds to the residues of organic matter generated from nutrient use within the body of water itself. This combination causes all stationary bodies of water to have short lives as measured by the geological time clock. Because they serve as sedimentation basins, they gradually fill, pass through a rooted aquatic stage as marsh land, and finally transform into ordinary land (Figure 4.5). The final stage is usually hastened by the digging of drainage ditches to make the land tillable, as evidenced by the great peat land farms in this country. In a few cases, such as in the Canadian Lakes in Michigan, marsh lands are dredged to recreate lakes.

Because of the ability of lakes and reservoirs to capture nutrient materials in the form of algae and rooted aquatics and to hold the residues of bacterial decomposition, it has often been said that "a major function of lakes and reservoirs is to hold on land areas those things that belong on the land." In surveys conducted between 1942 and 1944 on three Madison, Wis., lakes, retention of nitrogen was found to vary from 30% to as much as 60%.[4] Retention in ponds can approach 100%, although nitrogen can be lost from the system as N_2.

The considerations expressed above have led to a two-pronged program to preserve and prolong the life of our stationary bodies of water. The oldest of these is nutrient control mostly at point sources to prevent nuisance blooms. This program recently was broadened to include non-point sources. But only within recent times has nutrient control been considered an essential aspect of prolonging the life expectancy of lakes.

The second prong of the attack con-

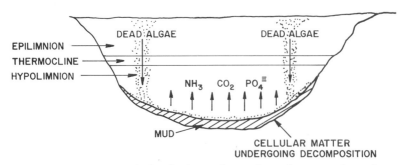

FIGURE 4.4. **Back-feed of nutrients from bottom deposits.**

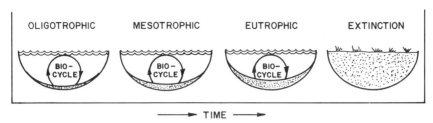

FIGURE 4.5. Natural transition of a lake through various stages of productivity, eventually resulting in extinction.

cerns the matter of silt deposition control. Silt is derived primarily from soil erosion and street washings. Air-borne silt is a minor consideration in most instances. Silt deposition is primarily of inorganic character and therefore permanently occupies space in a body of water. There is no way of removing it, except dredging. When Boulder Dam was built to create Lake Mead on the Colorado River, its life expectancy before being silted up was estimated to be only 150 years. Construction of Glen Canyon Dam and creation of Lake Powell upstream has greatly extended Lake Mead's expectancy because Glen Canyon will intercept much of the silt load.

The role of silt in enhancing the nutrient budget of a body of water is a matter of considerable debate. The debate centers over the ability of silt to release nutrients directly to waters above. Recent studies[5] have shown that silt has exchange properties of significance. At least the surface, and perhaps the entirety, of silt deposits are in equilibrium or strive to be in equilibrium with the minerals in the water above. When ammonia and phosphate concentrations are high in the water, silt tends to remove some by ion exchange. When the level is low in the water, it will release them. Thus, the silt functions as a nutrient "bank" or as a balancing device.

One aspect over which there is little difference of opinion concerns silt's role in providing needed nutrients to rooted aquatic plants. Growth of such plants, of course, results in production of organic matter containing some nutrients.

Upon death and decomposition, a portion of such nutrients is released to overlying waters.

Nitrogen versus phosphorus. The subject of nutrient control has had a very interesting history. From its beginning, attention was focused on control of phosphorus. This was the result of two factors. First, limnologists and agriculturists had long been aware of Liebig's Law of the Minimum, which, briefly stated, said, "The growth of any plant can be markedly curtailed by limiting the supply of one or more of the essential nutrients."

The second reason was related to the fact that among the essential elements— C, N, P, S, Ca, Mg, and K—only phosphorus (phosphates) seemed to have properties that permitted easy removal from wastewaters. Phosphorus in the form of phosphates and organic compounds could be easily removed by precipitation, lime, and coagulation or with multivalent cations from relatively inexpensive salts of aluminum or iron. Nitrogen removal was considered impractical because of the several forms in which it could occur and because of a lack of proven technology at the time. Very few municipalities chose to remove phosphorus because of the expense and problems of sludge disposal. Diversion of effluents away from critical waters was the accepted practice where possible.

In the late 1960s, renewed interest in removal of both nitrogen and phosphorus was stimulated by the Federal Water Quality Administration and more recently by that agency's successor, the U.S. Environmental Protection Agency. Through research based on federal grants, nitrogen removal was found to be possible by several methods:

- ammonia stripping with air,
- reduction of nitrates to nitrogen gas,
- ion exchange,
- break-point chlorination, and
- reverse osmosis.

Phosphorus removal was found to be feasible by addition of trivalent metal salts to activated sludge systems and, also, by changing activated sludge process operations to allow metabolic uptake of phosphorus, with no addition of chemicals.

By 1973, the way seemed clear for removal of both nitrogen and phosphorus at reasonable costs. But, the Arab oil embargo of 1974 and subsequent OPEC price increases have changed the economics of removal treatment so radically that serious doubts have arisen about the implementation for such expensive operations, particularly nitrogen removal.

The need for phosphorus removal has been accentuated to a considerable extent by the common use of phosphate-based detergents for household and industrial use. They have caused a marked increase in the phosphorus content of municipal wastewaters where their sale is permitted. This has changed the phosphorus:nitrogen ratio from about 1:8 to about 1:3 or 4. As a result, wastewaters today contain excessive amounts of phosphorus, far beyond the requirements of algae growths. This excess of phosphorus, after all the nitrogen has been used by the green algae, can be used by blue-green algae—*Anabaena, Aphanizomenon, Gleotrichia,* and *Nostoc*—that are capable of continued growth by fixing atmospheric nitrogen, which occurs in solution in all waters in plentiful supply. Figure 4.6 demonstrates in a simplified way how such action increases the nitrogen budget within a body of water. It constitutes an impor-

tant reason why control of phosphorus is absolutely essential.

LAND ENVIRONMENT

Wastewaters bearing nutrients are introduced to the land environment for any one of four reasons:

- irrigation water,
- nutrient removal,
- removal of BOD and suspended solids, or
- groundwater recharge.

Irrigation. Use of wastewaters for irrigation purposes has been widely practiced around the world, particularly in semi-arid or arid regions where rainfall is insufficient to interfere with land application as required by effluent quantities. In other areas, wastewater augments natural waters from pumped or stored supplies. In the U.S., it has been used on a wide variety of cultivated crops, pastures, and forest lands where disease transmission is not a factor.

Wastewaters from domestic sources are highly prized by agriculturists because of their nutrient content, primarily nitrogen, phosphorus, and potassium. A recent assessment at San Diego has placed nutrient values at about $0.025/m^3$ or about $31/ac-ft.

Wastewater has been used for irrigation purposes for over 50 years at Fresno, Calif., without apparant damage to the land. Numerous studies have indicated that crop, pasture, and forest lands effectively remove phosphorus and a large part of nitrogen from the water. The benefits derived from irrigation with wastewaters of domestic origin are great with respect to nutrient removal. This form of wastewater treatment constitutes the highest order of practice of returning to the land the things that came from the land.

Nutrient removal. In some instances, nutrient removal from wastewaters is necessary to protect receiving waters.

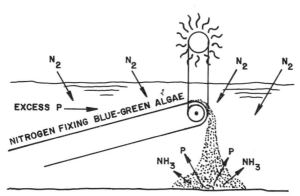

FIGURE 4.6. Diagram depicting how excess phosphorus increases nitrogen budget of natural waters and hastens eutrophication.

This can be accomplished by advanced or land treatment methods. Two notable cases where land treatment has been used for nutrient removal are State College, Penn., and Muskegon County, Mich.

At State College, serious degradation of Spring Creek had occurred (Figure 4.3). To relieve this situation, controlled experiments were conducted using the treatment plant effluent to irrigate crop and forest lands in the Spring Creek watershed.[6] Studies conducted on the groundwaters draining to Spring Creek from the experimental area showed them to be effectively free of phosphorus and a significant fraction of nitrogen. The nitrogen passing through the soil was converted to nitrates.

At Muskegon, which is located on the eastern shore of Lake Michigan, a large irrigation system was developed to remove nutrients from wastewaters as a means of protecting the lake. Because of location and climatic conditions, wastewater storage is needed for about 6 months flow. Percolate from the Muskegon County system is collected in underdrains and discharged into a nearby stream. Phosphorus removal has averaged 95 to 98% with percolate total phosphorus concentrations of 0.05 to 0.1 mg/L. Nitrogen concentrates in the applied wastewater has averaged 8.2 mg/L and the percolate total nitrogen has been 2.5 mg/L.

Groundwater recharge. In many areas of the world groundwater sources have been pumped out in excess of natural recharge. As a result, water levels have fallen and pumping costs have increased because of the greater lifts that must be overcome. In some cases, wells have gone dry and, in locations near the ocean, salt water intrusions have damaged aquifers. In some situations, such as in Peoria, Ill., recharge has been accomplished by diverting riverwater to recharge pits during the winter time when water temperatures are low. In this way, an adequate domestic water supply and suitable water for cooling purposes in industrial operations are kept available.

To correct the groundwater situation in many areas, particularly in semi-arid areas and in some densely populated places, such as Nassau County on Long Island, N.Y., projects involving recharge with wastewater treatment plant effluents have been initiated. Groundwater recharge is accomplished by either of two methods: deep well injection or surface spreading with resulting percolation.

Deep well injection. Deep well injection forces treatment plant effluents that must be treated to remove phosphates, nitrogen, and most suspended matter directly into aquifers. The Nassau County system and the Orange County,

Calif., Water Factory No. 21, are two plants designed to produce renovated wastewater suitable for well injection. In each case the primary purpose of injection is to arrest seawater intrusion.

Surface spreading and percolation. Where appropriately sloped land is available at reasonable cost, surface spreading and percolation often is the preferred method of groundwater recharge. Natural recharge, however, involves rainwater or run-off after contact with the soil. It is normally low in total dissolved solids (TDS) and carries little, if anything, that can be altered during passage through the soil. Because of the natural solvent powers of the water and carbon dioxide it can carry, natural water can dissolve considerable amounts of mineral matter while passing through soils. On the other hand, wastewaters are normally high in TDS and do contain both phosphates and nitrogenous compounds that can undergo significant changes in the soil.

One of the first large-scale facilities for groundwater recharge by surface spreading of wastewater effluent built in the U.S. was constructed in 1962 at the Whittier Narrows plant in Los Angeles County, Calif. Smaller installations of an experimental or permanent nature exist in other locations. Most of these have operated with a reasonable degree of success. However, stricter requirements for recharge are becoming increasingly evident.

REFERENCES

1. Birge, E. A., and Juday, C., "The Inland Lakes of Wisconsin. The Plankton I. Its Quantity and Chemical Composition." Wisconsin Geol. and Natural History Survey Bull. 64, Madison, Wis. (1922).
2. "Water Quality Criteria." Federal Water Quality Adm., Superintendant of Documents, Washington, D.C. (1972).
3. "Methodology for River-Quality Assessment with Application to the Willamette River Basin, Oregon." Geological Survey Circular 715M. U.S. Geol. Survey, Arlington, Va.
4. Sawyer, C. N., "Fertilization of Lakes by Agricultural and Urban Drainage." *J. N. Engl. Waterworks Assoc.*, **61,** 109 (1947).
5. To, Y.P.S., "The Effect of Sediment on Reservoir Water Quality." Ph.D. thesis, Va. Inst. of Technol., Civ. Eng. Dep., Blacksburg, Va. (1975).
6. Penneypacker, S. P., *et al.*, "Renovation of Wastewater Effluent by Irrigation of Forest Land." J. Water Pollut. Control Fed., **39,** 285 (1967).

Chapter 5

Biological Nitrification

PROCESS KINETICS

Microbiology. Biological nitrification in wastewater treatment systems is accomplished primarily by two genera of microorganisms. These genera, *Nitrosomonas* and *Nitrobacter,* are classified as autotrophic because they derive energy from inorganic compounds, as opposed to heterotrophic organisms that derive energy from organic compounds.

Nitrification of ammonium nitrogen is a two-step process. The first step, conversion of ammonium to nitrite, is mediated by *Nitrosomonas.* The second step, conversion of nitrite to nitrate, is mediated by *Nitrobacter* microorganisms. Equations 1 and 2 demonstrate this two-step conversion of ammonium to nitrate:

$$NH_4^+ + 3/2\ O_2 \rightarrow 2H^+ + H_2O + NO_2^- \quad (1)$$

$$NO_2^- + 1/2\ O_2 \rightarrow NO_3^- \quad (2)$$

Equations 1 and 2 are energy yielding reactions. *Nitrosomonas* and *Nitrobacter* use this energy for cellular growth and maintenance. In the above reactions, oxygen serves as the electron acceptor in the biochemical oxidation of both NH_4^+ and NO_2^-. Oxygen is the only electron acceptor that *Nitrobacter* or *Nitrosomonas* can use. An aerobic environment is therefore a necessary condition for nitrification to proceed.

Although the ammonium nitrogen in the nitrogen fraction converted by the genera *Nitrosomonas* and *Nitrobacter* is important, it is also important that the Total Kjeldahl Nitrogen (TKN) be identified, particularly if the wastewater has not undergone previous biological treatment. The NH_4^+-N/TKN ratio can vary widely even on domestic wastes because of wastewater temperature, collection line transport time, and type or lack of primary or preliminary treatment. In most plants the ratio of ammonium nitrogen:TKN increases in the summer because of sewer hydrolysis.

In the following discussions, ammonium nitrogen is used as the basis to calculate SRT (θ_c) and other levels. It should be understood, however, that TKN may better define the level of nitrogen that must be oxidized after correcting for nitrogen synthesized into the carbonaceous biomass.

When there are industrial wastes present, it is mandatory that both TKN and NH_4^+-N be evaluated comprehensively.

Stoichiometry. The energy produced by the oxidation of NH_4^+ and NO_2^- is used by nitrifying organisms primarily to produce new biomass (bacterial cells). These bacterial cells can be represented by the approximate chemical formulation, $C_5H_7O_2N$. The biomass synthesis reaction for *Nitrosomonas* and *Nitrobacter* is then given by Equation 3:

$$NH_4^+ + HCO_3^- + 4CO_2 + H_2O \rightarrow$$
$$C_5H_7O_2N + 5O_2 \quad (3)$$

It should be noted that bacterial cells in Equation 3 are produced entirely from inorganic compounds. It is assumed that micro-nutrients such as P, S, and Fe are necessary in small amounts for synthesis, but do not significantly alter Equation 3. This synthesis reaction requires an input of energy to proceed. During nitrification, this energy is obtained from NH_4^+ and NO_2^- oxidation (Equations 1 and 2). Reactions shown in Equations 1, 2, and 3, therefore, usually occur simultaneously. The energy yielded from the oxidation of one mole of NH_4^+ or NO_2^- is much less than the energy required to produce one "mole" of bacterial cells ($C_5H_7O_2N$). Equations 1, 2, and 3 then must be proportioned so that after energy transfer efficiencies are taken into account, energy use equals energy production.

Biological nitrification can be expressed by Equation 4:

$$NH_4^+ + 1.83 \ O_2 + 1.98 \ HCO_3^- \rightarrow$$
$$0.021 \ C_5H_7O_2N + 0.98 \ NO_3^-$$
$$+ 1.041 \ H_2O + 1.88 \ H_2CO_3 \quad (4)$$

Equation 4 can be used to estimate the three important stoichiometric parameters associated with the nitrification process: oxygen requirements, alkalinity consumption, and nitrifier biomass production. Equation 4 predicts that 1.83 moles of oxygen are required to convert one mole of ammonium into nitrate and new biomass. This is equivalent to 4.2 mg of O_2 per mg of NH_4^+-N converted. The overall equation predicts that 1.98 moles of H^+ are produced per mole of ammonium converted, or 7.14 mg of alkalinity (as $CaCO_3$) is destroyed per mg of NH_4^+-N converted. Finally the equation also predicts that 0.17 mg (0.021 "moles") of cells are produced per mg of NH_4^+-N converted.

The first two stoichiometric parameters—oxygen requirement and alkalinity destruction—are important in the design of nitrifying treatment systems. Oxygen is essential for the growth of nitrifiers, and alkalinity consumption is important because nitrifying organisms are efficient only over a relatively narrow pH range. The third stoichiometric parameter, biomass production, is important to the understanding of microbial population dynamics. The low cell production frequently causes washout of nitrifiers in combined carbon-nitrogen oxidation systems without high sludge age. Compared to heterotrophic oxidation of carbonaceous waste material, nitrification generates relatively little biomass. Consideration of these stoichiometric parameters will be demonstrated in the design example at the end of this section.

Nitrification kinetics. Engineering design of nitrification systems must take into account kinetic relationships, as well as the stoichiometric relationships discussed above. Kinetic relationships essentially describe the rate at which the reactions proceed. The rate of ammonium oxidation is directly proportional to the growth rate of the nitrifying organisms. The most widely accepted approach for describing the growth kinetics of microorganisms is by the use of the Monod expression (Equation 5):

$$\mu = \frac{\hat{\mu}S}{K_s + S} \qquad (5)$$

where

μ = specific growth rate of microorganisms, day^{-1},

$\hat{\mu}$ = maximum specific growth rate of microorganisms, day^{-1},

K_s = half velocity constant, which is the equal of the growth limiting substrate concentration at half the maximum growth rate, mg/L, and

S = concentration of growth limiting substrate, mg/L.

Excluding O_2 and micronutrients, the growth limiting substrate for *Nitrosomonas* is usually NH_4^+. The growth limiting substrate for *Nitrobacter* is usually NO_2^-. The growth rate of *Nitrobacter* is much greater than the growth rate of *Nitrosomonas* for a wide variety of conditions. Because the nitrification process is sequential (that is, $NH_4^+ \rightarrow NO_2^- \rightarrow NO_3^-$), the overall nitrification process is usually rate controlled by the *Nitrosomonas* reaction. Hence, the overall process kinetics can effectively be described by *Nitrosomonas* growth kinetics.

The Monod expression for nitrification then takes the following form (Equation 6):

$$\mu_N = \frac{\hat{\mu}_N N}{N + K_N} \qquad (6)$$

where

N = NH_4^+-N concentration, mg/L, and

K_N = half-saturation constant, mg/L of NH_4^+-N.

The rate of ammonium oxidation is thus controlled by the growth rate of *Nitrosomonas*, and is related to this growth rate by the yield coefficient as expressed in Equation 7:

$$q_N = \frac{\mu_N}{Y_N} \qquad (7)$$

where

q_N = ammonium oxidation conversion rate, g NH_4^+-N oxidized/g of *Nitrosomonas*, day^{-1}, and

Y_N = yield coefficient, g *Nitrosomonas* produced/g NH_4^+-N removed. It should be noted that the "NH_4^+-N removed" includes oxidation plus assimilation into new biomass.

The biological solids retention time, θ_c, is defined as the ratio of the total biomass in the system divided by the total amount of biomass leaving the system daily. In some biological wastewater systems, the nitrifier biomass is only a relatively small fraction of the total system biomass. The total system biomass typically is dominated by heterotrophic organisms. As long as the nitrifier biomass fraction is constant in both the total system biomass and in the total removed biomass, θ_c for the nitrifiers equals θ_c for the total (nitrifiers + heterotrophs) biomass.

At steady state, the amount of biomass leaving the system daily equals the amount of daily biomass growth. Thus, θ_c and μ_N are related as shown in Equation 8:

$$\theta_c = 1/\mu_N \qquad (8)$$

Environmental factors that significantly influence the rate of nitrification are dissolved oxygen (DO) concentration, pH, temperature, and the presence of inhibitors. Each of these factors is considered in the following sections.

In the design of nitrification systems, the engineer will find that the definition of design parameters in fixed film systems lacks much of the quantification that is available for suspended film systems. However, it is suspected that the fixed film systems will respond in a similar manner to pH, temperature, DO, and inhibitory substances. Documentation is lacking and difficult if not impossible to define in the manner used for suspended film reactors.

Effects of dissolved oxygen. The presence of DO is an essential condition for

27

nitrification. Several investigators have proposed that as the DO concentration decreases, DO (rather than NH_4^+) might become the growth limiting factor. It has been proposed that when DO becomes growth limiting, the Monod expression can be modified as follows (Equation 9):

$$\mu_{N_{DO}} = \hat{\mu}_N \left(\frac{DO}{K_{O_2} + DO} \right) \qquad (9)$$

where

DO = dissolved oxygen concentration, mg/L,

K_{O_2} = half saturation constant for oxygen, mg/L, and

$\mu_{N_{DO}} = \mu_N$ corrected for limiting DO.

Experimental evidence as summarized in the EPA Technology Transfer Manual on Nitrogen Control[2] shows a considerable range in observed K_{O_2}. This range, 0.15 to 2.0 mg/L, is so large that it is questionable whether any value can be assumed for use in Equation 9. More studies need to be conducted to clearly define K_{O_2}. In the meantime, the DO effect should be recognized. To avoid having DO limit nitrification rates, systems should be designed and operated so that the DO level should never be allowed to drop below 1 mg/L. However, a DO concentration of 2 mg/L or higher offers a greater margin of safety, and no correction for DO is required.

Effects of pH. The effect of pH on nitrification is significant. Many studies have shown that there is a narrow optimum pH range for nitrifier growth. Most of these studies indicate that this optimum range is somewhere between pH 7.5 and 8.6. There have been attempts to quantify the effect of pH on nitrification rates. However, variability in experimental results makes quantification questionable.

For design and operation purposes, it is sufficient to take into consideration that nitrification performance drops off rapidly as pH is lowered. However, nitrification does not seem to be severely limited up to pH 9.5. For nitrification performance stability, it is best to maintain pH on the basic side of neutrality (above pH 7.2).

These pH considerations are important because nitrification consumes alkalinity. If the bicarbonate alkalinity remaining after NH_4^+-N oxidation is less than 50 mg/L (as $CaCO_3$), provision for alkalinity addition is required. Peak ammonium conditions must be evaluated. If adequate alkalinity is not present to buffer the system, the pH will drop and nitrification will be inhibited. The design example at the end of this section demonstrates this situation.

Effects of temperature. As in most biochemical reactions, temperature has a great influence on nitrification rates. The problem of quantifying the temperature effect is very difficult, as evidenced by the widely varying observations in the literature. For design and operational purposes, it is more meaningful to use reasonable estimates, based on experimental evidence of $\hat{\mu}_N$ and K_N at various temperatures. Table 5.1 presents conservative estimates of the maximum growth rates of *Nitrosomonas* at three temperatures, 10°, 20°, and 30°C. These values were taken from the literature as summarized in the EPA manual for nitrogen control.[2] The values in this table agree reasonably well with the van't Hoff-Arrhenius relationship, which predicts doubling of rates with each 10°C increment in temperature.

For wastewater temperatures between 10° and 20°C, a linear extrapolation of $\hat{\mu}_{N_{10}}$ of 0.3/day and $\hat{\mu}_{N_{20}}$ of 0.65/day is adequate. A similar interpretation can be made between 20 and 30°C using $\hat{\mu}_N$ of

TABLE 5.1. Estimates of $\hat{\mu}_N$ for *Nitrosomonas*.

Temperature (°C)	$\hat{\mu}_N$(days^{-1})
10	0.3
20	0.65
30	1.2

0.65 and 1.2/day, respectively. For temperatures above 30°C, a $\hat{\mu}_N$ of 1.2/day should be used unless pilot data are available. For temperatures less than 7°C, $\hat{\mu}_N$ is difficult to predict without pilot studies.

Experimental data indicate that K_N increases with increases in temperature. Experimental data, collected mostly for the temperature range 15° to 25°C, suggest that K_N has a value in the range of 0.5 and 3.5 mg/L NH_4^+-N. However, the data are so inconsistent that it appears that assuming a value of K_N equal to 1 mg/L of NH_4^+-N will be as useful, for design purposes, as trying to precisely define K_N.

The K_N and $\hat{\mu}_N$ values in the range of 5° to 9°C are highly unpredictable. The design engineer should fully evaluate the impact of a higher discharge of ammonium nitrogen during cold weather. High removals may not be cost effective nor necessary.

Effects of inhibitors. Nitrifying organisms are very sensitive and are susceptible to a relatively large array of inorganic and organic inhibitors. Quantification of inhibition effects by these compounds has not been accomplished to date. These large numbers and types of inhibitory agents preclude listing and discussion here. Instead, the comprehensive review by Painter[3] should be consulted. Ammonia and nitrous acid in high concentrations also can be inhibitory to nitrification. A detailed discussion of this is given by Anthonisen et al.[4]

DESIGN OF NITRIFICATION SYSTEMS USING KINETIC APPROACH

The single most important design parameter in designing a nitrifying system is the growth rate of the nitrifying organisms. Selection of the appropriate design θ_c is based on the degree of nitrification required. The NH_4^+-N remaining in the effluent is directly related to θ_c by

Equations 6 and 8. These equations can be combined to give Equation 10:

$$\theta_c = \frac{K_N + N}{\hat{\mu}_N N} \quad (10)$$

The design θ_c from Equation 10 usually is increased by a factor of 1.5 to 2.5 to allow the system a margin of safety to accommodate typical fluctuations in flow, ammonium nitrogen loading, and environmental factors. Once θ_c has been selected, estimates can be made for oxygen and alkalinity requirements.

Unlike much of the primary effluent BOD_5, peak ammonium nitrogen levels cannot be absorbed by the biological floc for later conversion. Because no storage is possible, residual ammonium nitrogen levels are more dependent on the instantaneous NH_4^+-N quantity to the nitrifier mass ratio corrected for the detention characteristics of the aeration basin or the fixed film reactor. For this reason, the rate of TKN and ammonium entering the system and the ability to maintain biological mass are important design criteria. The rate of ammonium nitrogen as TKN entering the system should be evaluated as a function of the retention time of the nitrification system. There is little reason to consider maximum hourly TKN rate for an extended aeration system with 24-hour detention. However, it will be very significant in the design of a second-stage nitrification system of 4-hour detention or a trickling filter of 5-minute retention.

For these reasons, the mass flow rate of TKN used for design should represent the type and final design of the nitrification system used.

The following example demonstrates the general use of growth kinetics in the design of a nitrification system. Design of suspended growth (combined and separate systems) and fixed growth reactors are described in detail later in this chapter.

Sample problem. To demonstrate the general growth rate design procedure, it is assumed that a wastewater with 20 mg/L NH_4^+-N is to be nitrified in a biological system so that the effluent contains no more than 1 mg/L of NH_4^+-N. The lowest expected temperature of the wastewater is 20°C. Alkalinity of the wastewater is 250 mg/L as $CaCO_3$. The pH of the waste is 7.5. The preliminary design for a system is as follows:

At 20°C, $\hat{\mu}_N$ is estimated (from Table 5.1) to be 0.65 day^{-1}. K_N (for all temperatures) is estimated to be 1.0 mg/L NH_4^+-N. If N equals 1 mg/L, then Equation 10 predicts that θ_c is equal to 3.1 days. A safety factor (described in a following section) of 2 gives a design θ_c = 6.2 days.

Oxygen consumption can be calculated (Equation 4) to be 4.2 mg/L per mg/L of NH_4^+-N converted to new cells and nitrate. In this case, 20 minus 1, or 19 mg/L of NH_4^+-N, are converted. Therefore, the oxygen required is 4.2 mg/L $O_2 \times$ 19 mg/L NH_4^+, or 79.8 mg O_2/L. Alkalinity consumption can be calculated in a similar fashion using Equation 4. In this case, alkalinity consumption equals 7.14 mg alkalinity consumed per mg NH_4^+-N used. Therefore, alkalinity destroyed equals 7.14 $\times$ 19 mg/L, or 135.66 mg/L as $CaCO_3$. It seems that in this case there is adequate alkalinity to buffer the system, although the pH will drop slightly.

Discussion of design. Several important design considerations have been neglected in the above preliminary design. First, there is no consideration made of aeration tank size, that is, hydraulic detention time. Aeration tank size essentially determines the mixed liquor biomass levels and the required oxygenation rate. Sizing of the aeration unit is complicated by other factors, which are discussed in a following section. Aeration volume is defined by the concentration of suspended solids in the basin which may be limited by the clarification/thickening capacity. Another important consideration that has been neglected is the difference between completely mixed and plug flow aeration units. The above discussion applies mainly to completely mixed systems. However, use of the above approach for plug flow systems requires only a modification of Equation 10 to yield Equation 11:

$$\frac{1}{\theta_c} = \frac{\mu_N(N_0 - N_1)}{(N_0 - N_1) - K_N \ln \dfrac{N_0}{N_1}} \quad (11)$$

where

N_0 = influent ammonium concentration, mg/L NH_4^+-N, and

N_1 = effluent ammonium concentration, mg/L NH_4^+-N.

The accuracy of this design approach can be demonstrated in Figure 5.1, which compares predicted effluent quality (NH_4^+-N) with actual operating data from nitrifying wastewater treatment plants.

Finally, the preliminary design procedure outlined above assumes that the wastewater contains no organic matter, and therefore, the aeration basins lack heterotrophic activity. If heterotrophic activity is present, as in a combined carbon oxidation-nitrification system, then both oxygen demand and the required mass of mixed liquor suspended solids will be increased.

Because the heterotrophic biomass yield, Y_H, is relatively high, most of the biomass in a combined system is heterotrophic. An estimate can be made of the portion of the mixed liquor biomass that is autotrophic. If Y_H is defined as grams of heterotrophic biomass produced per gram of ultimate BOD (BOD_L) removed, then the amount of heterotrophic biomass produced per day is approximately QS_LY_H. S_L equals the influent concentration of BOD_L and Q equals the influent flow rate. Estimating heterotrophic biomass production in this manner assumes that the endogenous decay often associated with heterotrophic activity is negligible. This estimate also assumes complete removal of BOD_L. Nitrifier biomass production can be estimated in a similar manner. Nitrifier biomass production can be estimated by QN_oY_N. A

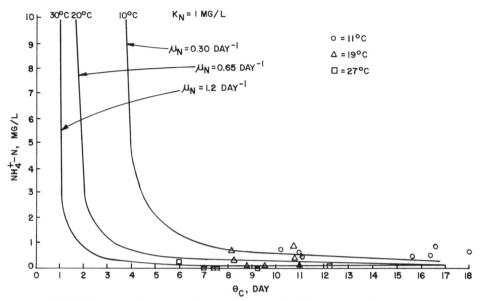

FIGURE 5.1. Predicted and actual nitrification performance.

ratio (f) of autotrophic biomass to total biomass is given in Equation 12:

$$f = \frac{Y_N \cdot N_o}{(Y_N \cdot N_o) + (S_L \cdot Y_H)} \qquad (12)$$

Typical values of Y_N, N_o, Y_H, and S_L used in Equation 12 demonstrate that f is usually very small. For example if $Y_N = 0.15$, $Y_H = 0.4$, $Y_N = 250$ mg/L, and $N_o = 25$ mg/L, then $f = 0.036$.

In a combined system, oxygen demand for heterotrophic activity must be added to that for autotrophic activity (Equation 4). An estimate of this additional (heterotrophic) oxygen can be made if it is assumed that the portion of organic matter (BOD_L) that is converted to heterotrophic biomass does not exert an oxygen demand. If heterotrophic biomass is represented as $C_5H_7O_2N$, then this biomass exerts 1.42 g of BOD_L per gram of biomass. In other words, each gram of heterotrophic biomass produced represents 1.42 g of BOD_L removed, but not oxidized. Therefore, heterotrophic oxygen demand is estimated by multiplying the BOD_L removed by the factor $(1 - Y_H \cdot 1.42)$. For example, if BOD_L removal is equated to $Q \cdot S_L$ (an approximation used earlier), heterotrophic oxygen demand can be estimated to be $Q \cdot S_L$ (1–0.4 (1.42)). It should be noted that heterotrophic oxygen demand as expressed here has units of milligrams of oxygen per day, if Q is expressed as L/day. Autotrophic oxygen demand represented in the same units is $Q \cdot N_o$ (4.2).

NITRIFICATION IN SUSPENDED-GROWTH SYSTEMS

Biological nitrification is applicable in those cases where an ammonium removal requirement is mandated without need for complete nitrogen removal. It is also the first step in biological nitrification/denitrification.

Nitrification may be achieved by suspended-growth or attached-growth processes.[1,2,5-7] Suspended-growth processes are those that suspend the nitrifiers in a mixed liquor by some mixing mechanism, while attached-growth processes maintain the bulk of the nitrifiers on the media. Representative nitrification processes are listed in Table 5.2.

31

TABLE 5.2. Classification of nitrification facilities.

Type and location	Scale mgd	BOD$_5$/TKN ratio	Oxygen demand distribution in percentage		Ref.	Classification degree of separation		Pretreatment to remove BOD$_5$
			BOD$_5$	NOD		Combined oxidation-nitrification	Separate stage	
Suspended Growth								
Manassas, Va.	0.2	1.2	20	80	1		×	Activated sludge
Hyperion, Los Angeles, Calif.	46	7.3[a]	61	39[a]	2	×		Primary treatment
Central Contra Costa Sanitary Dist., Calif.	1.0, design 30	2.4	34	66	3		×	Lime primary treatment
Livermore, Calif.	pilot	1.0	18	82	4		×	Activated sludge
Flint, Mich.	3.3	2.8	38	62	5		×	Roughing filter
Valley Community Services District, Calif.	34	5.5	65	35[a]	6	×		Primary treatment
Blue Plains, Washington, D.C.	3.8	10.8[a]	70	30[a]	3	×		Primary treatment
	pilot, design 309	1.3 to 3.0	22 to 39	61 to 78	7,8		×	Activated sludge
Whittier Narrows, LACSD, Calif.	12	6.6	61	39	9	×		Primary treatment
Jackson, Mich.	13.5	9	66	34	10	×		Primary treatment
Tampa, Fla.	pilot design 60	3.0	40	60	11		×	Activated sludge
South Bend, Ind.	pilot	1.8	28	72	12		×	Activated sludge
New Market, Ontario, Canada	2.4	2.6	36	64	13		×	Lime primary treatment
Cincinnati, Ohio	pilot	7.2	61	39	14	×		Primary treatment
	pilot	1.0[b]	18[b]	82[b]	15		×	Activated sludge
Fitchburg, Mass.	pilot	1.0	18	82	16		×	Activated sludge
Marlboro, Mass.	pilot	3.6[c]	40	60	17		×	Trickling filter
Amherst, N.Y.	pilot	0.8 to 2.0	22	78	18		×	Activated sludge
Denver, Colo.	pilot	2.7	37	63	19,20		×	Activated sludge
Attached Growth								
Stockton, Calif.	pilot design 58	5.3	54	46	21	×		Primary treatment
Midland, Mich.	pilot	1.1	19	81	22		×	Trickling filter
Union City, Calif.	pilot	1.7	27	73	23,24		×	Activated sludge
Allentown, Pa.	40	1.9	30	70	25		×	Trickling filter
Lima, Ohio	pilot	0.79	15	85	26,27		×	Activated sludge

[a] Approximate, calculated from COD data.
[b] Calculated from effluent.
[c] BOD·NH$_3$-N ratio; BOD/TKN would be about 3.0.

BOD$_5$:TKN ratio is one of the critical factors for nitrification processes. Early attempts at nitrification combined carbon oxidation and nitrification as occur in the extended aeration modification of the activated sludge process. Because of high ratios of BOD$_5$:TKN in such systems, the nitrifier population level is generally low, and the majority (60 to 70%) of the oxygen requirements are a result of organic carbon demand.

Where nitrification is achieved in a separate stage, BOD$_5$ ratio to influent ammonium load is low, resulting in the presence of a higher nitrifier population and subsequently higher rates of nitrification. Most of the oxygen requirements in this case are caused by ammonium oxidation. This nitrogen oxygen demand or requirement will be referred to as NOD.

In Table 5.2, if the BOD$_5$:TKN ratio is 4 or lower, the nitrification system is arbitrarily classified as a separate stage nitrification process. If the BOD$_5$:TKN ratio is greater than 4, the process is classified as a combined carbon oxidation-nitrification process.

Table 5.2 also shows the distribution of oxygen demand between BOD$_5$ and NOD. For separate nitrification processes, NOD accounts for at least 60% of the total oxygen demand.

COMBINED SUSPENDED GROWTH NITRIFICATION PROCESSES

Carbon oxidation-nitrification in suspended-growth reactors. There are several variations of combined carbon oxidation-nitrification systems used in suspended growth reactors. Six such variations of the activated sludge processes are described in the following text.

Complete mix modification of the activated sludge process. The feed and withdrawal arrangement of a complete mix plant is shown in Figure 5.2. The complete mix design provides uniformity of load to all points within the reactor, thereby easing the problem of oxygen dissolution at the head end of a

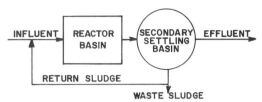

FIGURE 5.2. Complete mix and extended aeration plants.

reactor of conventional plants. Because the total basin will be at the equilibrium pH, this design may be more sensitive to loss of alkalinity during ammonium oxidation than a plug flow system.

Extended aeration plants. Extended aeration plants are similar to complete mix plants except that the hydraulic detention times range from 24 to 48 hours rather than 2 to 8 hours. Endogenous respiration is maximized in the extended aeration plants and solids retention times of 30 days can be expected.

Conventional or plug flow plants. The hydraulics of the conventional system approach the plug flow configuration because the influent wastewater and return activated sludge are returned to the head end of the reactor and the combined flow passes along a long narrow tank before exiting the system. Minimum pH occurs in the latter stages or length of the tank during nitrification.

Contact stabilization plants. In the contact stabilization process the return activated sludge is separately aerated in a sludge reaeration tank prior to mixing with the influent wastewater as shown in Figure 5.3.

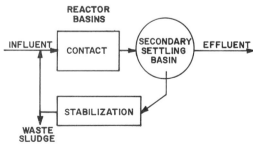

FIGURE 5.3. Contact stabilization plant.

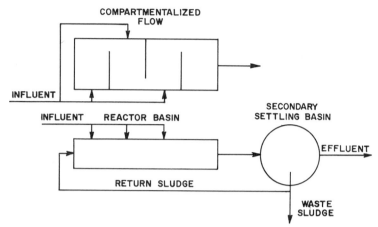

FIGURE 5.4.   Step aeration plant.

BOD$_5$ removal takes place in the contact tank, which has a short detention time of 0.5 to 1 hour. The removal is attributed to adsorption of particulate and colloidal solids by the biological solids. Only partial nitrification can be obtained in this process because of an insufficient number of nitrifiers in the biological mass. Because peak hydraulic and organic loads would produce very unfavorable conditions for nitrification, this concept is discouraged. Further, high BOD$_5$ removals required to produce nitrifying conditions often cannot be produced.

The complete design procedure is found in the EPA Nitrogen Control manual.[4] Operating plant results were not described in the literature.

Step aeration plants. In the step aeration plants, the return sludge is introduced at the head end of the aeration tank and the influent wastewater is in-

troduced at several points along the aeration tank, as shown in Figure 5.4. A variation on step aeration known as sludge reaeration does not introduce influent in the first pass of the aeration tank, thus establishing a sludge reaeration zone. Only partial nitrification is obtained because of insufficient contact time for organic nitrogen conversion to ammonium nitrogen and ammonium oxidation. Like contact stabilization, this flow sheet does not offer the advantages of complete mix, plug flow, or orbital systems for nitrification.

Orbital aeration systems. Various modifications of the Passveer oxidation ditch are more frequently being considered for nitrification (and nitrification-denitrification) (Figure 5.5). These systems have a very high recirculation rate, yet are neither complete mix nor plug flow (from a biological kinetics standpoint). The systems are aerated-pro-

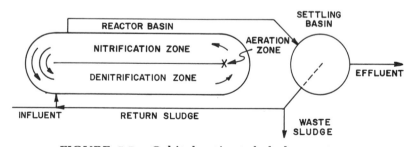

FIGURE 5.5.   Orbital activated sludge system.

pelled by rotor, disk, mechanical, jet, or propeller pump-diffuser devices.

The alkalinity needs are often lower for orbital systems because the process can be operated in the partial denitrifying mode during lower flows and during warmer periods when $\hat{\mu}_N$ is higher. This is the most common nitrification concept used in Europe.

The aeration volume for this "low load" system is similar to that for extended aeration.

Design of combined suspended-growth nitrification reactors based on solids retention time and nitrification rates. Nitrification kinetic theory can be applied to define the following design parameters:

• The safety factor required to assure no significant ammonium level in the effluent during peak flows. The safety factor depends on the nitrification system used, and specifically on retention time in the nitrification system;

• The solid retention time under most adverse conditions of pH, DO, temperature, and SVI. Inhibitory compounds within the sludge are not considered in the design, as they should be eliminated from the system as part of industrial pretreatment;

• The allowable organic loading on the combined carbon oxidation nitrification stage;

• The hydraulic detention time and tankage configuration; and

• Excess sludge wasting schedule.

The procedure is termed the "solids retention time," SRT (θ_c), design approach.

The following example describes the design considerations and magnitudes for a complete mix activated sludge system.

Design example. A 10-mgd (440 L/s) treatment plant is required to achieve complete nitrification at 15°C. Primary effluent BOD$_5$ is 150 mg/L and total Kjeldahl nitrogen (TKN) is 25 mg/L and NH$_4$-N is 20 mg/L as N. The wastewater alkalinity is 280 mg/L as CaCO$_3$. Determine the SRT and, therefore, the design criteria. Complete nitrification is assumed where residual NH$_4$-N is equal to or less then 1.0 mg/L.

Solution:

1. Establish the safety factor (SF). SF is affected by the desired effluent quality and the ratio of the maximum TKN to the average TKN loading and the aeration volume or detention. The minimum SF is 1.5 and may be increased depending on the particular conditions of each case. In this case, an SF of 2.0 is selected since it is a medium-size plant with moderate peak/average loadings.

2. Select the minimum mixed liquor dissolved oxygen concentration, DO. A minimum DO level of 2.0 mg/L is recommended to avoid depressing effects of low DO on the rate of nitrification. This is to be maintained at peak BOD$_5$ and TKN loadings.

3. Determine the process operating pH. It is recommended that a pH level of 7.2 or higher be maintained. Every mg/L of NH$_4^+$-N oxidized causes the destruction of 7.14 mg/L of alkalinity as CaCO$_3$. The alkalinity remaining after nitrification is $280 - (25 \times 7.14)$ or 102 mg/L. No significant change in pH is anticipated.

4. Calculate the maximum growth rate of nitrifiers at 15°C, DO of 2 mg/L, and pH > 7.2:

$$\hat{\mu}_{N_{15}} = \hat{\mu}_{N_{10}} + \frac{20 - t}{20 - 10} (\hat{\mu}_{N_{20}} - \hat{\mu}_{N_{10}}) \quad (13)$$

where

$\hat{\mu}_{N_{15}}$ = maximum nitrifier growth rate, day^{-1}, for pH $\geq$ 7.2, temperature of 15°C and DO $\geq$ 2 mg/L,

$\hat{\mu}_{N_{10}}$ and $\hat{\mu}_{N_{20}}$ = maximum nitrifier growth rate, day^{-1}, temperatures 10 and 20°C, and

t = operating temperature, 15°C.

Substituting into Equation 13:

$$\hat{\mu}_{N_{15}} = 0.30 + \frac{5}{10} (0.65 - 0.30)$$

$$\hat{\mu}_{N_{15}} = 0.475 \text{ d}^{-1}$$

5. Calculate the minimum SRT, θ_c^m, for 15°C operation, $K_N = 1$, NH_4^+-N = 1.0 mg/L. From Equation 10:

$$\theta_c^m = \frac{K_N + N}{\hat{\mu}_N N}$$

$$\theta_c^m = \frac{1.0 + 1.0}{(0.475)(1.0)} = 4.2 \text{ d} \qquad (14)$$

6. Select a safety factor (1.5 to 2.5) and the corresponding design solids retention time, θ_c^d:

$$\theta_c^d = \theta_c^m \times SF \qquad (15)$$

Let SF be 2.0: $\theta_c^d = 4.2 \times 2.0 = 8.4$ d.

7. Derive the design nitrifier growth rate:

$$\mu_N = \frac{1}{\theta_c^d} = \frac{1}{8.4} = 0.12 \text{ d}^{-1} \text{ at } 15°C$$

8. Check the design nitrifier growth rate (step 7) against Equation 6:

$$\mu_{N_{is}} = \frac{\hat{\mu}_N N}{N + K_N} = \frac{(0.475)(1)}{1.0 + 1.0}$$

$$\mu_{N_{is}} = 0.238 \text{ d}^{-1} \qquad (16)$$

The growth rate determined from Equation 6 (the Monod expression) is considered conservative. It compares favorably with the design nitrifier growth rate, μ_N, of 0.12 d^{-1} from Step 7.

9. Calculate the organic removal rate, recognizing that θ_c^d applies to both nitrifiers and heterotrophic populations:

$$\mu_b = \frac{1}{\theta_c^d} = Y_b q_b - K_d \qquad (17)$$

where

Y_b = heterotrophic yield coefficient, 0.65 kg VSS synthesized/kg of BOD$_5$ removed (0.65 lb VSS/lb BOD$_5$),

q_b = daily rate of substrate BOD$_5$ to VSS removal, kg BOD$_5$ removed/kg VSS·d,

K_d = endogenous decay coefficient, 0.05 d^{-1}, and

μ_b = heterotrophic growth rate, d^{-1}.

For representative values for Y_b of 0.65 and K_d of 0.05:

$$\frac{1}{8.4} \text{ d} = 0.65 \, q_b - 0.05 \text{ d}^{-1}$$

q_b = 0.26 kg BOD$_5$ removed/kg MLVSS·d

10. Determine the hydraulic detention time. For this determination, the MLVSS content and effluent soluble BOD must be known. The effluent soluble BOD can be assumed to be very low, or in the range of 2 mg/L. If the MLSS, X_1, is assumed to be 2500 mg/L and 75% volatile, then the hydraulic detention time, t, is:

$$\bar{t} = \frac{S_o - S_e}{x_b q_b} \qquad (18)$$

where

$\bar{t}$ = hydraulic detention time, d,

x_b = MLVSS, mg/L,

S_o = influent total BOD$_5$, mg/L, and

S_e = effluent soluble BOD$_5$, mg/L.

$$\bar{t} = \frac{150 - 2}{(0.75 \times 2\,500)(0.26)}$$

$$\bar{t} = 0.304 \text{ d}$$

$$\bar{t} = 7.3 \text{ h}$$

11. Determine the organic loading per unit volume

$$V = Q\bar{t} \qquad (19)$$

where

V = volume, L (mil gal),

$V = (440 \text{ L/s})(0.304 \text{ d})$

$V = 11.2 \text{ ML, } (3.04 \text{ mil gal})$

$V = 11\,200 \text{ m}^3 \text{ (406 400 cu ft)}.$

Organic loading, lb BOD$_5$/day/1000 cu ft, is:

$$BOD_5 = Q S_o \qquad (20)$$

36

where

S_o = BOD$_5$ of the influent, mg/L,
BOD$_5$ = (10 mgd) (150 mg/L) (8.34),
BOD$_5$ = 12 490 lb/day,
BOD$_5$ load/1000 cu ft =
$$\frac{12\,490}{406\,400}\,(1\,000)$$
= 30.8 lb BOD$_5$/day · 1000 cu ft.

13. Determine the waste sludge quantity:

$$WAS = Y - TSS_e \qquad (21)$$

where

WAS = waste activated sludge, lb TSS/day,
Y = net yield, lb TSS/day, and
TSS_e = effluent TSS, lb/day.

Net yield, Y, is:

$$Y = \frac{I}{\theta_c^d} = \frac{(V)(X_1)(8.34)}{\theta_c^d} \qquad (22)$$

where

I = MLSS under aeration, lb.

Therefore

$Y =$
$$\frac{(3.04 \text{ mil gal})(2\,500 \text{ mg/L})(8.34 \text{ lb/gal})}{8.4 \text{ days}}$$
= 7550 lb TSS/day, and
I = 63 420 lb MLSS.

With an effluent TSS of 15 mg/L, the waste activated sludge is:

WAS = 7550 − (10 mil gal)(15 mg/L)
(8.34 lb/gal)
= 6300 lb TSS/day
= 4720 lb VSS/day (2140 kg/d)

In this example, a safety factor of 2.0 was used. In small plants with high peak/average flows, BOD$_5$ and TKN loadings, a higher factor may be appropriate. If a safety factor of 2.5 had been used in the above example, the θ_c^m would be 10.5 days, with suitable adjustment made in aeration volume.

Graphical Design Procedure. An alternative graphical design procedure can be used to obtain the nitrification aeration tank volume for combined carbonaceous-nitrification systems. This procedure is based on the solids retention time design approach and the relationship between aeration tank volume, mixed liquor solids concentration, net solids yield, amount of BOD removed, flow and design solids retention time as follows:[8]

$$V = \frac{Y_t\,(\Delta BOD_5)Q\,\theta_c^D}{X} \qquad (23)$$

where:

V = aeration volume, m^3,
ΔBOD_5 = BOD$_5$ removed, mg/L,
Q = influent flow, m^3/d,
X = aeration tank mixed liquor total suspended solids, mg/L,
θ_c^D = design solids retention time, days, and
Y_t = net solids yield coefficient, kg. TSS/kg BOD removed.

Values for the net solids yield vary as a function of solids retention time, endogenous respiration rate, and influent inert solids. Values used in this graphical procedure are within the range of those described elsewhere[1] for aeration systems without primary treatment or following primary treatment. Inert or non-biodegradable solids are present at a higher concentration for influents without primary treatment resulting in a higher net solids coefficient. The design solids retention is the product of the solids retention time predicted by Equation 10 and the design safety factor. The design safety factor is the Peak/Average ammonia loading ratio. Figure 5.6 shows the design solids retention time as a function of the activated sludge temperature and safety factor.

Rearranging Equation 23 results in the following relationship used to develop Figures 5.7 and 5.8 for nitrification of activated sludge treating either raw wastewater or primary effluent:

$$\left(\frac{X}{\Delta BOD}\right)\left(\frac{V}{Q}\right) = Y_t\,\theta_c^D \qquad (24)$$

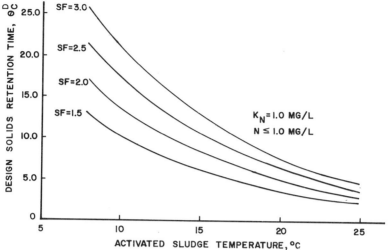

FIGURE 5.6. Nitrification system design solids retention time as a function of temperature and the safety factor.

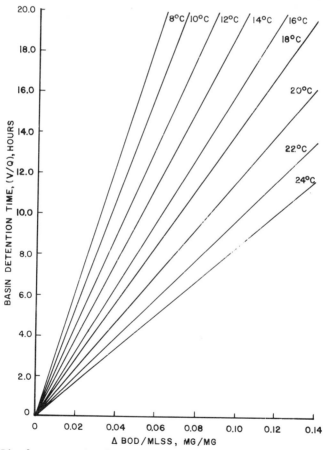

FIGURE 5.7. Single stage nitrification design curves for an activated sludge system treating raw wastewater. The safety factor is equal to 2.0.

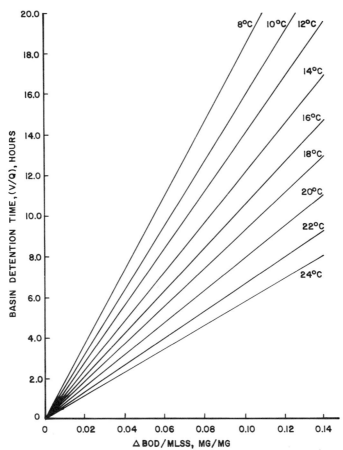

FIGURE 5.8. Single stage nitrification design curves for an activated sludge system treating primary treated wastewater. The safety factor is equal to 2.0.

The nitrification basin detention time is determined from the ratio of the design BOD removal and mixed liquor suspended solids concentrations. For both design curves a safety factor of 2.0 is assumed. The basin detention time for other design safety factors is determined by multiplying by the selected design safety factor, divided by 2.0.

Design Example. The same design conditions as assumed in the previous design example will be used:

$$\text{Flow} = 10 \text{ mgd}$$
$$(440 \text{ L/s})$$
$$\text{Primary Effluent BOD}_5 = 150 \text{ mg/L}$$
$$\text{TKN} = 25 \text{ mg/L}$$
$$\text{Design Safety Factor} = 2.0$$
$$\text{Temperature} = 15°C$$
$$\text{Assumed MLSS} = 2500 \text{ mg/L}$$

Solution:

1. Design the aeration system to maintain a mixed liquor dissolved oxygen concentration of 2.0 mg/L and assure that sufficient alkalinity is present to maintain a pH of 7.0 to 7.2.

2. Obtain the basin detention time:

$$\Delta\text{BOD} = 150 - 2 = 148 \text{ mg/L}$$
$$\Delta\text{BOD/MLSS} = 148/2500 = 0.059$$

From Figure 5-8; Detention time = 6.6 hr (0.28 days).

3. Basin Volume

$$\text{Volume} = (440 \text{ L/s}) (0.28 \text{ days})$$
$$= 10.4 \text{ ML } (2.75 \text{ mil gal})$$
$$= 10\,400 \text{ m}^3 (367\,000 \text{ cu ft}).$$

39

4. BOD Loading

$QS_o/V = (1 \text{ mgd}) (150) (8.34)/36.7$
$= 34 \text{ lb BOD/day-1000ft}^3$

5. Sludge Wasting

$$\text{Sludge Production} = \frac{XV}{\theta_c^D}$$

From Figure 5.7, $\theta_c^D = 8.5$ days
Therefore,

Sludge Prod. $= (2500) (2.75) (8.34)/8.5$
$= 6750 \text{ lb/day TSS}$

Effluent Solids

$= (10 \text{ mg/L}) (10) (8.34)$
$= 834 \text{ lb/day}$

Net sludge wasting:

WAS $= 6750 - 834$
$= 5916 \text{ lb/day TSS}$
@ 75% Volatile $= 0.75 (5916)$
$= 4437 \text{ lb VSS/day}$

The aeration tank volume and sludge production values are within 10% of the values determined by the previous de-

sign calculation method. The differences are due to assumptions used for sludge yield coefficients. However, in both designs values for θ_c^D are identical.

If no primary treatment were used for the same design example, Figure 5.7 would be used to determine the basin detention time. A time of 10.5 hours is necessary, assuming the same aeration tank MLSS concentration. This is due to the fact that for the same solids retention time a higher net solids yield exists because of the presence of more nonbiodegradable solids in the raw wastewater than in primary effluent. These solids should improve settling, so that MLSS concentration of 3 000 mg/L could be used, resulting in a basin detention time of 8.8 hours.

Extended Aeration Design. The above graphical design procedure can be used to design extended aeration systems including orbital activated sludge systems. The basin detention time design curves are shown in Figure 5.9. Design

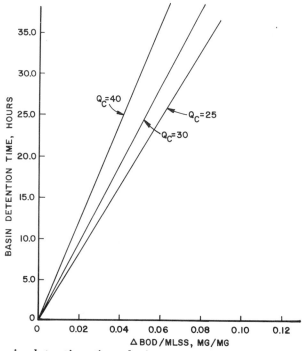

FIGURE 5.9. Basin detention time design curve for an extended aeration system treating raw wastewater.

solids retention times ranging from 25 to 40 days used for extended aeration designs are normally adequate for nitrification, BOD₅ removal, and sludge stabilization with wastewater temperatures down 7° to 8°C. The previously described design procedure can be used to determine the aeration tank volume and sludge production as a function of the design solids retention time.

Discussion of effects of temperature and safety factor on the complete mix design. Table 5.3 shows the effects of various temperatures and safety factors on the calculated design parameters for a 44 L/s (1-mgd) complete mix activated sludge plant.

Selection of a safety factor in plant design is generally practiced by choosing design solids retention time θ_c^d, greater then θ_c^m. A minimum SF of 1.5 is required for complete nitrification in the effluent, that is, a residual concentration of 0.5 to 2.0 mg/L NH₄-N. Safety factor curves for plug flow and complete mix activated sludge plants operating at 10° and 20°C are shown in Figures 5.10 and 5.11.

Safety factor determinations depend on factors such as:

● Desired degree of nitrification,
● Transient load conditions,
● Presence of industrial waste in the influent, and
● Ratio of maximum TKN loading to the average loading. The maximum loading is computed for a period equal to the aeration detention time, for example, peaking factor/average load equal to 1.0 for 24-hour aeration.

As a rule of thumb, the minimum safety factor should equal or exceed the ratio of peak ammonium nitrogen load to average load to prevent high ammonium nitrogen escape in the effluent at peak loads.

A safety factor equal to or less than 1.25 is suggested if θ_c^D is equal to or greater than 25 days.

Design considerations for other suspended growth nitrification reactors.

Extended Aeration Kinetics—The same kinetics as apply to the complete mix process are also applicable to the extended aeration process. The safety factor can be low because of the long retention times.

Conventional Activated Sludge Kinetics—Substrate removal rates for plug flow tanks are integrated over the period of time an element of liquid remains in the nitrification tank. The removal rate equation for plug flow tanks is reduced to the following expression for a return sludge:influent ratio of less than one:

$$\frac{1}{SF} = \frac{(N_0 - N_1)}{(N_0 - N_1) + K_N \ln \dfrac{N_0}{N_1}} \quad (25)$$

Plug flow is more efficient for the same safety factor than complete mix, thus requiring less aeration volume for the same level of nitrification efficiency. However, plug flow reactors suffer from high oxygen demand requirement at the head end of the tank, and special air diffusion system must be considered to handle this demand. If oxygen supply limitations are present at the head end of the tank, the plug flow advantage over the complete mix reactor is lost.

Step Aeration Activated Sludge Kinetics—The step feed pattern of step aeration plants causes the kinetics of such plants to approach more closely the complete mix process. The design approach for complete mix can be used for step aeration plants yielding reasonable results.

Orbital Flow Activated Sludge Kinetics—Because of the long detention times of these systems, their kinetics are similar to extended aeration. To obtain full use of basin volume, the DO returning to the aeration device should equal or exceed 1.0 mg/L. Lower DO can be used when the oxygen uptake is low without affecting nitrification. These systems offer high reliability due to ease of con-

TABLE 5.3. Calculated design parameters for a 44 L/S (1-mgd) complete mix activated sludge plant. Simultaneous BOD₅ and NOD removal.

Minimum temp. for nitrification, °C	Maximum possible nitrifier growth rate μ_N, day^{-1}	Assumed allowable MLSS/MLVSS mg/L	Safety factor SF	Design solids retention time, days θ_c^d	Steady-state effluent NH_4^+-N mg/L	Organic removal rate lb BOD rem/ lb MLVSS-day	Hydraulic retention time, hr	BOD-loading (volumetric) lb/1000 cu ft/day
10	0.300	2,000 / 1,500	2.0	13.3	<1.0	0.19	12.5	18.0
			2.5	16.7	<1.0	0.17	13.9	16.2
			3.0	20.0	<1.0	0.15	15.8	14.2
15	0.475	2,500 / 1,875	2.0	8.4	<1.0	0.26	7.3	30.8
			2.5	10.5	<1.0	0.22	8.6	26.1
			3.0	12.6	<1.0	0.20	9.5	23.7
20	0.65	3,000 / 2,250	2.0	6.2	<1.0	0.33	4.8	46.8
			2.5	7.7	<1.0	0.28	5.6	40.2
			3.0	9.2	<1.0	0.24	6.6	34.1

Note: Table developed from the following data;

$K_N = 1.0$ mg/L NH_4^+-N
$N = 1.0$ mg/L NH_4^+-N
$Y_b = 0.65$ lb VSS/lb BOD₅ removed
$K_d = 0.05$ day^{-1}
$S_o = 148$ mg TBOD₅/L influent
$S_e = 2$ mg SBOD₅/L effluent

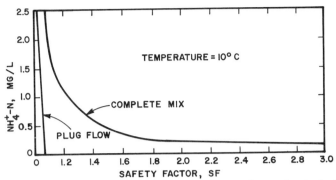

FIGURE 5.10. Safety factor curve for a complete mix activated sludge plant operating at 10°C.

trol and low rate of change in operating conditions. Partial denitrification and alkalinity recovery is often achieved.

SEPARATE STAGE SUSPENDED GROWTH REACTORS

Pretreatment. Nitrification in suspended growth reactors may be accomplished in separate stage reactors as well as in the previously described combined reactors. The influent BOD_5:TKN ratio to separate stage nitrification typically ranges from 1:1 to 3:1. This level of carbon removal is greater than can be accomplished by primary treatment alone. Pretreatment methods include chemical treatment, activated sludge,

roughing filters, trickling filters, or rotating contactors followed by clarification.

The pretreatment used greatly influences the performance of the separate stage nitrification process. The following are some of the effects of various pretreatment methods.

Pretreatment by chemical addition. Chemicals may cause changes in the alkalinity and pH of the influent to the separate stage nitrification units. Loss of 5.6 mg of hydroxyl alkalinity occurs for each mg of Al^{+3} as alum added. There is also a loss of 3.7 mg of phosphate alkalinity for each mg of Al^{+3} used for phosphorus removal. The effect of

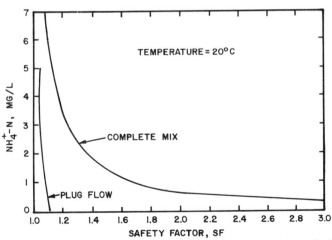

FIGURE 5.11. Safety factor curve for a complete mix activated sludge plant operating at 20°C.

this loss of alkalinity on the separate stage nitrification units must be considered if insufficient alkalinity is available to maintain adequate pH.

Lime addition generally raises the pH of the influent to the separate stage nitrification reactor and changes the level of alkalinity based on the lime dose and the quality of the raw wastewater. High pH values of the wastewater enhance nitrification, but generally are lowered by recarbonation prior to discharge to receiving waters. Carbon dioxide from the oxidation of organic carbon and acidity from the nitrification reactions greatly reduces or eliminates the need for recarbonation by carbon dioxide addition.

Concern over lack of phosphorus for nitrification in the biological secondary as a result of chemical primary treatment may be unwarranted because the phosphorus requirement for nitrification is very low; in the range of 0.1 to 0.5 mg/L.

Degree of organic carbon removal. Low levels of organic carbon in the nitrification process influent are advantageous to attached growth reactors because these low levels may eliminate the need for clarification following the nitrification process. In suspended growth systems, low total organic carbon in the influent may cause an imbalance between the solids lost from the sedimentation units and the solids synthesized in the reactor. This condition necessitates continuous wasting or increasing of BOD in the influent to maintain the inventory of biological solids in the nitrification system.

Protection against toxicants. Toxic metals are best treated by chemical primary treatment using lime. Organic toxicants are not affected by chemical treatment unless coupled with carbon adsorption. Biological pretreatment methods provide a degree of protection from toxicants except for organics resistant to biological oxidation.

DESIGN OF SEPARATE STAGE SUSPENDED GROWTH REACTORS

Nitrification kinetics. The nitrification rate is the accepted design approach for separate stage nitrification. Experimentally measured rates are more appropriate for use than are theoretical nitrification rates because of the difficulty of assessing the nitrifier fraction of the mixed liquor.

The solids retention time design approach is also applicable for separate stage nitrification. The design approach is dependent on determining the net solids production for the system. This is a function of the second stage influent solids, influent BOD, and ammonia oxidized. Because the solids yield of nitrifying bacteria is relatively low, this can be ignored in determining the system MLSS concentration, net solids yield, and solids retention time. The main concern in the design of a second step nitrification system is the possibility of sludge production being less than that which normally can be lost from the system as effluent solids. When this possibility exists the separate stage approach is not recommended unless additional BOD_5 or solids can be fed to the separate stage. Normally the separate stage suspended film nitrification will follow a previous biological stage that produces a BOD_5 effluent of 40 mg/L or more. It is not cost effective to design for separate stage suspended film nitrification of secondary level effluent quality.

Solids Retention Time Approach—Graphical Procedure. In this design approach the aeration tank MLSS concentration and solids production must be based on the influent solids and the influent BOD_5 concentration caused by the previous step. A portion of the influent solids will be biological solids and will undergo endogenous respiration in the second stage aeration. The other solids are considered inert in this design procedure. The following equation, similar to Equation 23, describes the MLSS concentration as a function of the influent solids:

$$\frac{V}{Q} = \frac{F\,(X_e)\,\theta_c^D}{(1 + b\theta_c^D)\,X} + \frac{(1 - F)\,(X_e)\,\theta_c^D}{X}$$

(26)

Where:

F = Fraction of influent volatile suspended solids,

X_e = Influent suspended solids to second stage nitrification aeration tank, mg/L,

$\frac{V}{Q}$ = Aeration tank detention time, days, and

b = Endogenous respiration coefficient, day^{-1}

The aeration time required to maintain a design solids retention time due to soluble BOD removal must be added to that determined by Equation 24. This aeration time can be determined from Figure 5.8 as a function of the design temperature, BOD removed, and MLSS concentration.

Figure 5.12 can be used to determine the aeration time described in Equation 26. This is based on an endogenous respiration coefficient of b equal to 0.08 day^{-1} and an assumption that F is 0.8. If other values are desired, then Equation 26 can be used to calculate this aeration time requirement. Figure 5.6 is used first to determine the design solids re-

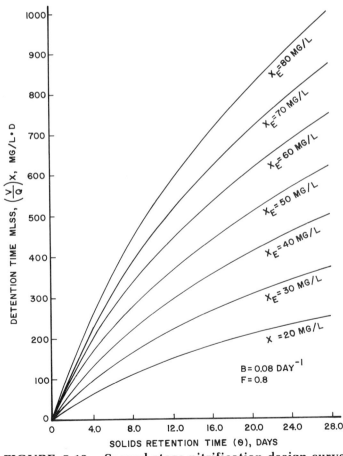

FIGURE 5.12. Second stage nitrification design curve.

tention time for Figure 5.12. Based on assumed design values for influent solids (X_e) and MLSS concentration (X), the basin detention time is determined. This detention time is added to that determined from Figure 5.8 based on BOD removal.

Design Example.
Assume the following trickling filter effluent for feed to a nitrification aeration tank:

$$\text{Flow} = 1 \text{ mgd } (44 \text{ L/S})$$
$$\text{Soluble BOD}_5 = 20 \text{ mg/L}$$
$$\text{Suspended Solids} = 40 \text{ mg/L}$$
$$\text{Temperature} = 15°C$$
Safety Factor = 2.0 (1st stage dampens peaks)
$$\text{Effluent NH}_4\text{-N} = 1.0 \text{ mg/L}$$
Basin MLSS = 2000 mg/L (low because of difficulty of solids separation and thickening)

Solution:

1. Determine the design solids retention time

$$\theta_c^D = 8.5 \text{ days (Figure 5.6)}$$

2. Determine aeration time due to BOD removal

BOD = 20 − 2 = 18 mg/L
Δ BOD/MLSS = 18/2000 = 0.009
Detention time; $t_1 = 1.0$ hr (Figure 5.8)

3. Determine aeration time due to influent solids

$$X_e = 40 \text{ mg/L}$$
$$\frac{V}{Q}(X) = 228 \text{ mg/L·days @ } \theta_c^D$$
$$= 8.5 \text{ days (Figure 5.12)}$$
$$\frac{V}{Q} = \frac{(24)(228)}{2000}$$
$$t_2 = 2.7 \text{ hr}$$

4. Total detention time at MLSS = 2000 mg/L

$$t_1 + t_2 = 1.0 + 2.7$$
$$= 3.7 \text{ hr } (0.15 \text{ days})$$

5. Aeration tank volume

Volume = (44 L/S) (0.15 days)
$$= 0.57 \text{ ML } (0.15 \text{ mil gal})$$
$$= 568 \text{ m}^3 (20{,}050 \text{ cu ft})$$

6. Sludge Wasting

Sludge Production = $(X)(V)/\theta_c^D$
$$= 2000 (0.15) 0.34/8.5$$
$$= 294 \text{ lb/day TSS}$$
Effluent Solids = (20)(8.34) 1
$$= 167 \text{ lb/day TSS}$$
WAS = 294 − 167
$$= 127 \text{ lb/day}$$
@ 80% volatile = 102 lb/day VSS

A significant problem with a second stage nitrification system is that it is necessary to minimize effluent solids losses to build up an adequate MLSS concentration and nitrification population in the aeration tank. This design example illustrates the importance of influent solids and BOD to generate an adequate MLSS in the aeration basin.

Nitrification rate (r_N) approach. Design using nitrification depends on experimentally determined rates from pilot studies. As expected, nitrification rates increase as the temperature increases. The BOD_5:TKN ratio strongly influences the nitrification rates with nitrification rates increasing as the ratio decreases. Nitrification rates are also effected by the pH of the mixed liquor and how far it deviates from the optimum pH for nitrification. Figure 5.13 displays some observed nitrification rates at various locations.

Peak nitrification rates are expressed as:

$$r_N = q_N \cdot f \qquad (27)$$

where

r_N = peak nitrification rate, g NH_4^+-N oxidized/g MLVSS/day,
f = nitrifier fraction, and
q_N = ammonia oxidation rate, g NH_4^+-N oxidized/g nitrifiers/day

The rates derived from Figure 5.13 are for the condition where ammonium level and DO are not limiting. The DO effect

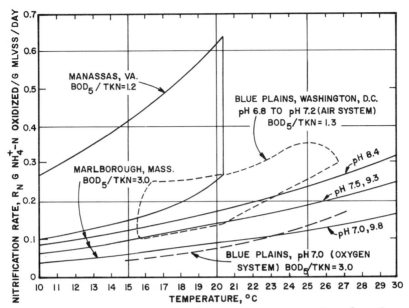

FIGURE 5.13. Observed nitrification rates at various locations.

and ammonium level are taken into consideration using the correction for DO and SF. The design nitrification rate (from Equation 9) is expressed as:

$$r_N = \frac{r_N}{SF}\left(\frac{DO}{K_{O_2} + DO}\right)$$

where the term $(DO/K_{O_2} + DO)$ approaches 1.0 when DO is greater than 2.0 mg/L. The effluent ammonium content (from Equations 9 and 15) is estimated using the following expression:

$$\frac{1}{\theta_c^m SF} = \frac{N \mu_N}{N + K_N}$$

where $K_N = 1.0$

Aeration requirements. Sufficient oxygen must be provided to handle the load impressed on the nitrification process at all times. Diurnal variations in ammonium load have to be determined carefully so that the aeration equipment can be designed with sufficient capacity to handle variations during peak ammonia load.

In addition to peaking of ammonium or organic nitrogen, a concurrent peak of organic substance may occur. This situation must be considered in the selection of aeration equipment. Flow equalization equipment may be considered as an alternative to extra tankage and aeration capacity.

Proper design dictates that careful consideration be given to maximizing oxygen use per unit power input. In the face of significant load variation, the aeration system should be designed to match the load variation while economizing on power input.

Mechanical surface aerators are less suited for nitrification applications because they normally are designed to operate at fixed speeds. Diffused air aeration has the advantage of being easily modulated to match the load by turning down or switching off individual blowers. Diffusers can be arranged for even distribution of energy input in the reactor. Gentler mixing is also achieved in diffused air system than in mechanical aeration system. This may produce lower effluent TSS except in the case of orbital systems, where low effluent solids have been demonstrated with surface aeration.

47

BIOLOGICAL NITRIFICATION

Oxygen dissolution requirement. Oxygen requirements in a nitrification system may be presented as:

$$OD = f\,BOD_5 + NOD \qquad (28)$$

where

OD = total oxygen demand, mg/L,
f = multiplier, generally 1.2-to-1.3, depending on SRT and temperature, and
NOD = oxygen required to oxidize TKN, taken as 4.6 times the TKN removed less the TKN contained in the excess cell mass, mg/L.

Temperature correction and correction factors relating wastewater characteristics must be applied to the rated performance of the aerator as supplied by the manufacturer.

The heterotrophic oxygen demand (f) of BOD_5 as a function of the system θ_c is provided in the ASCE-WPCF Treatment Plant Design manual, Figures 14-17 and 14-18. Temperature effects between 10° and 30°C are noted on Figures 14-17.

pH control. In cases where the alkalinity of the water is depleted by nitrification, supplemental alkalinity by chemical addition must be considered. Upstream processes using alum or iron reduce the wastewater alkalinity and this effect should also be considered.

Chemical addition. Lime is the preferred chemical to add for low pH control because of its availability and cost. In suspended growth systems operated in the plug flow mode, probes may be positioned at several points in the aeration tank with provisions for chemical addition at several points also. In systems operated in the complete mix mode, chemical addition is controlled by monitoring effluent pH.

Effect of aeration method on chemical requirements. Carbon dioxide is produced at the rate of approximately 1.4 kg for each kg of oxygen used for the oxidation of BOD_5 and NOD. The effects of oxygen transfer efficiency and residual

TABLE 5.4. Effect of oxygen transfer efficiency and residual alkalinity on operating pH.

Residual alkalinity as CaCO₃ mg/L	pH at Stated transfer efficiency, %			
	6	9	12	18
50	6.9	6.7	6.6	6.5
75	7.1	6.9	6.8	6.7
100	7.2	7.0	6.9	6.8
125	7.3	7.1	7.0	6.9
150	7.4	7.2	7.1	7.0
175	7.4	7.3	7.2	7.0
200	7.5	7.3	7.2	7.1

alkalinity on operating pH are shown in Table 5.4. Diffused air and pure oxygen aeration tend to suppress the pH to a greater extent than mechanical aeration. Plug flow aeration has minimum CO_2 pH suppression at the end of the basin where nitrification pH suppression is the highest (where there is the highest level of nitrates).

NITRIFICATION IN ATTACHED-GROWTH SYSTEMS

In addition to suspended growth systems, there are a variety of alternative biological processes currently available to nitrify domestic and industrial wastewaters. Collectively they are referred to as attached-growth or fixed-film systems. In these systems biomass is attached to solid support media contained within a reaction vessel. Although the kinetic (growth and oxidation) relationships used in design of suspended growth reactors are still valid, design of attached-growth systems is complicated by mass transfer limitations of substrate into the biological films (Figure 5.14) and definition of the film surface itself. Because of this, design of attached-growth reactors is primarily based on empirical results from pilot- and full-scale systems.

Examples of attached-growth systems include trickling filters, rotating biological contracors, packed bed reactors,

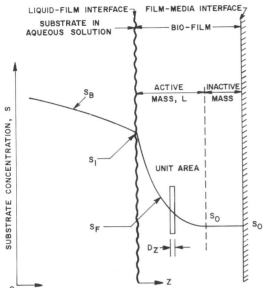

FIGURE 5.14. Heterogenous system composed of substrate in aqueous solution and microbial population in biological film.

and fluidized beds. In general, attached-growth systems may tend to be more resistant to shock BOD loads and, when enclosed, are protected from excess losses in temperature. However, the short detention time in some fixed-film reactors may result in breakthrough of ammonium nitrogen at peak mass flows.

Examination of Figure 5.14 shows that the substrate typically penetrates only a portion of the biofilm. Only a portion of the biomass is therefore actively involved in metabolizing the substrate and the ratio of active to inactive mass is related to the thickness of the biofilm. An additional advantage of attached-growth systems is that there is no need for recirculated sludge to maintain the necessary biomass to accomplish treatment because the biomass is attached to the solid support in the reactor. From a practical design viewpoint, this means process efficiency is not dependent on the settleability of the biomass as is the case for suspended-growth systems.

As with the suspended-growth systems, nitrification may be accomplished in a separate unit process or in combination with carbonaceous BOD removal. Both modes of operation will be examined.

The design engineer must consider the peak ammonium loadings, as well as the average loading rate. One method is to design the plant for average conditions and then increase the reactor size in proportion to the peak to average ammonium loading rates. In cases where the peak : average ratio is very high, the use of flow equalization may be appropriate, or high θ_c suspended growth reactors may be more economical and efficient.

Trickling filters. The operation of trickling filters is well known and is discussed in detail in other manuals of practice.[1,5] Basically, the process consists of distributing the incoming wastewater through a rotary distributer or a nozzle system, and allowing it to trickle or flow downward through a bed of stone, wood, or plastic media that is covered with a film of microorganisms. As the wastewater flows downward, it absorbs oxygen which, together with the organic matter in the wastewater, is removed by the bacterial biofilm.

Trickling filters in the early 1980s are being used much more extensively for removal of ammonia. The application of trickling filters for upgrading existing plants appears to be equally divided between use as a roughing filter prior to the activated sludge process and as a tertiary nitrification unit following a previous biological step and clarification. An equal number of projects are using trickling filters in parallel or series to achieve ammonium nitrogen removal simultaneously with carbon oxidation. Economy of power and space as well as ease of operation seem to be the primary reasons for selecting trickling filters.

Systems using rock media generally are designed between 1.2 and 3.0 m (4 and 10 ft) in height. Towers using redwood or plastic media may be designed up to 9 m (30 ft) in height because of the relative light weight and greater void space of the media. Because of this and because they generally have 100 to 300% more surface area per volume, plastic and redwood media filters generally require less space than conventional stone filters.

Loading capacity is known to be a function of the available surface area for biological growth, organic loading, temperature, pH, dissolved oxygen, and the presence of toxic or inhibitory substances. A measure commonly used to gauge the available surface area is specific surface, the amount of clean media surface available per unit volume of the reactor. It is usually expressed in terms of (m^2/m^3) or (ft^2/ft^3). Use of media with specific surfaces greater than 100 m^2/m^3 (31 ft^2/ft^3) has been avoided for simultaneous BOD_5 and NOD removal to reduce the potential problem of clogging the reactor media with excessive biological growth.

Combined carbon oxidation-nitrification. The effect of organic loading on nitrification efficiency of rock media has been summarized by EPA in Figure 5.15. Organic loading has been seen to affect nitrification efficiency because the bacterial film in the rock is dominated by heterotrophic bacteria at high organic loadings.

For rock media, attainment of 75% nitrification or better requires organic loading to be limited to 0.16 to 0.19 $kg/m^3 \cdot d$ (10 to 12 lb BOD_5/day/1000 cu ft). At higher organic loading rates the degree of nitrification diminishes. Above 0.48 to 0.64 $kg/m^3 \cdot d$ (30 to 40 lb BOD_5/day/1000 cu ft), very little nitrification is observed.

The greater the specific surface, the greater the volumetric quantity of active microorganisms. Therefore, when higher

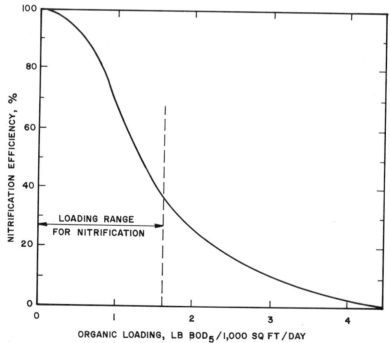

FIGURE 5.15.   Combined carbon oxidation-nitrification performance. BOD loading for plastic media was 40 lb/1000 cu ft/day (840 g/m$^3 \cdot$ d); for rock, 25 lb/1000 cu ft/day (400 g/m$^3 \cdot$ d).

specific surface plastic media is used, higher organic loadings can be applied while still achieving good nitrification. It has been demonstrated[7] that plastic media with a specific surface of 88.5 m^2/m^3 (27 sq ft/cu ft) could be loaded at 400.5 $g/m^3 \cdot d$ (25 lb/day/1000 cu ft) and achieve complete nitrification. Another factor that may allow the plastic media to be loaded at higher rates is that plastic media filters normally have better ventilation and therefore better oxygen transfer. One cautionary note is in order: plugging of the media by heterotrophic growths may occur if specific surface areas greater than 98.4 m^2/m^3 (30 sq ft/cu ft) are used.

Trickling filter data for rock and plastic can be correlated if the data are plotted on a basis of BOD_5 loading of the surface area of the media, $kg/m^2 \cdot d$ (lb/day/1000 sq ft) of BOD_5. This relationship (Fig. 5.15) indicates that plastic media with 80% higher surface area will be able to nitrify about 60% more $NH_4^+ \cdot N$ per unit volume in a combined BOD_5 and NOD removal system.

The beneficial effect of recirculation on nitrification is thought to be primarily caused by increased oxygen supply and reduced NH_4^+-N:DO ratio. To ensure uniform wetting of the plastic media filters, a minimum hydraulic application rate of 0.47 to 0.68 $L/m^2 \cdot s$ (0.7 to 1.0 gpm/sq ft) is required. Without recirculation, nitrification design loadings may result in inadequate hydraulic loadings. Recirculation is therefore important in preventing portions of the media from drying out, ensuring reasonable flow distribution and reducing ammonium:DO ratios.

The roughing filter-activated sludge sequence is often referred to as a dual biological system. In the roughing mode, it will generally remove 60 to 80% of the BOD on a settled basis and 40 to 65% unsettled. In many cases roughing treatment of the primary effluent prior to an existing aeration basin results in nitrification in the existing basin without further modification.

An important advantage of the dual biological system is the ability to control SVI to the range of 60 to 100 cm^3/g on municipal wastes. This allows the activated sludge to operate at higher MLSS because of better settlement and improve the volumetric nitrification efficiency of the aeration basin.

Separate stage nitrification. The volumetric rate of nitrification (mass removed/unit volume of reactor/unit of time) is proportional to the available surface area for biological growth. The films normally associated with nitrification are much thinner than for combined carbonaceous BOD and ammonium removal. As a result, media with higher specific surfaces may be used without increasing the probability of plugging and ponding occurring. Specific surfaces range from 88.5 to 223 m^2/m^3 (27 to 68 sq ft/cu ft) for the dumped- and corrugated-sheet module type medias.

The most extensive data to date on nitrification with trickling filters has been generated at the Midland, Mich., and Lima, Ohio, pilot facilities. A summary of results is graphically displayed in Figure 5.16. It should be noted that an increase or decrease in organic loading on the plant will affect the required surface area. As can be seen, reducing effluent ammonium concentrations below 3.0 mg/L requires significant increases in reactor size. There is inadequate data to define the effects of peak NH_4^+-N loadings on the trickling filter. Because the liquid retention time in a plastic media filter is 3 to 5 minutes, this system will "see" instantaneous peak NH_4^+-N loadings.

Historically, ammonium loadings to rock filters have been based on reactor volume rather than surface area, primarily because the specific surface of the rock media was relatively constant and not a design factor. In general, rock filters exhibit the same general trends as do plastic filters—higher organic loadings or decreased temperatures require

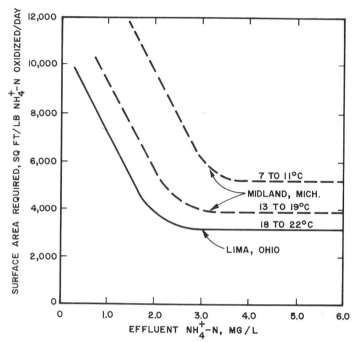

FIGURE 5.16. **Separate stage nitrification for Midland, Mich., and Lima, Ohio.**
NOTE: sq ft/lb $\times$ 0.2 = m^2/kg.

increased reactor volumes to achieve the desired effluent quality.

Recirculation apparently has marginal effect on nitrification efficiency on an average basis. However, greater consistency has been reported when recirculation has been employed. A recycle ratio of 1:1 is considered adequate for dry weather flows for most applications.

Rotating biological contactors. The rotating biological contactor (RBC) is a fixed-film biological wastewater treatment system. The equipment consists of large-diameter corrugated plastic media mounted on a horizontal shaft and placed in a reactor tank. The media is slowly rotated while approximately 40% of the surface area is submerged in the wastewater. Immediately after start-up, organisms naturally present in the wastewater begin to adhere to the rotating surfaces and multiply until, in about one week, the entire surface is covered with up to an approximately 1.6-mm thick layer of biomass for carbonaceous BOD$_5$ removal and a 0.8-mm thick layer

for nitrification. This large microbial population achieves high degrees of treatment for relatively short wastewater detention times. The growth has a shaggy appearance and contains a mixed biological population. The growth provides a large active biological surface area—much larger than the surface area of the media. In rotation, the media carries the organisms and a film of wastewater into the air at a peripheral speed of 0.15 to 0.30 m/s. In this portion of the rotational arc, oxygen quickly diffuses into the film and the biological layer. The organisms in the biomass remove organic materials from the film of wastewater, using oxygen diffused directly from the atmosphere for their respiration. Further removal of dissolved oxygen and organic materials occurs as the media continues rotation through the bulk of wastewater in the tank. Residual dissolved oxygen in the film is diffused into the mixed liquor, thus maintaining a mixed liquor dissolved oxygen concentration.

Shearing forces exerted on the biomass as it passes through the waste-water strip excess biomass from the media into the mixed liquor. This maintains a uniform biomass thickness and eliminates media clogging. The use of an additional air supply has shown thinner biomass films than strictly mechanical systems because of the additional shear provided by bubbles passing over the surface of the media. The mixing action of the rotating media keeps the stripped solids in suspension until the flow of treated wastewater transports them out of the process for separation and disposal.

During the first week or so of operation, a shaggy biomass structure develops on the media. The shaggy structure of the biomass is not caused by the presence of any particular species of organism, but by the action of rotation. The continual drag from being rotated through the wastewater, and the draining of entrained wastewater when rotated up into the air, causes the growth to form elongated macroscopic filaments. This "shaggy" structure is most apparent in the initial stages of the media, where the BOD concentrations are higher and the growth is thicker, and gradually decreases in the later stages. Unlike the trickling filter, rotating contactors use controlled hydraulic shear, that is, "rock-speed," as the principal mechanism for

displacing excess biological growth. The shaggy growth will increase in length and thickness until it can no longer withstand the forces exerted by rotation. The stripped growth is in the form of relatively large aggregates of dense biomass that settle rapidly in a final clarifier.

Figure 5.17 shows a process flow diagram for a treatment plant incorporating the rotating biological reactors. Raw wastewater flows first through primary treatment for removal of large objects and floatable and settleable materials. Primary effluent then flows to a multistage biological reactor where fixed aerobic cultures or microorganisms absorb the dissolved organic matter from the wastewater and flocculate the suspended matter for subsequent gravity separation.

Each stage of media operates as a completely mixed, fixed-film biological reactor in which there is a dynamic equilibrium between the rate of biological growth and the rate of stripping biomass. Treated wastewater and stripped biomass pass through each subsequent shaft of media. As wastewater passes from stage to stage, it undergoes a progressively increasing degree of treatment by the different biological cultures that develop in the successive stages.

Initial stages of media, which receive

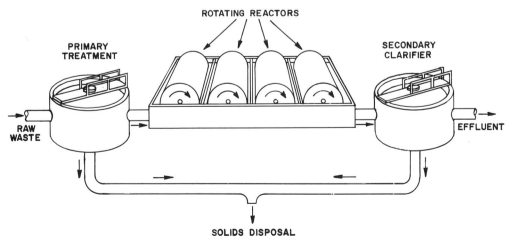

FIGURE 5.17. Process flow diagram for rotating biological reactors.

the highest concentration of organic matter, develop cultures of mixed heterotrophic bacteria. As the concentration of organic matter decreases in subsequent stages, nitrifying bacteria begin to appear along with various types of higher life forms, such as rotifiers, vorticella, and other predators.

Stripped biomass and flocculated suspended matter leave the last stage of media with the treated wastewater, and pass to a secondary clarifier where the solids are separated for disposal. Clarified effluent passes on for disinfection or further treatment. Settled solids thicken to 1.5 to 3% solids in the secondary clarifier. When secondary solids are recycled to the primary clarifier (when treating domestic wastewater), a combined sludge of 3 to 5% solids is normally produced for disposal.

Stripped reactor solids retain their density in the mixed liquor and settle rapidly in the secondary clarifier. Mixed liquor suspended solids concentration flowing to the final clarifier is approximately equal to the influent suspended solids concentration, which will typically vary from 50 to 200 mg/L when treating domestic wastewater. The process is operated on a once-through basis with no need for recycle of sludge or effluent, unless there are dissolved oxygen limitations in the first stage(s). Splitting the flow to the first two stages can eliminate oxygen deficiencies.

Nitrification potential. In recent years, numerous researchers have investigated the kinetics of nitrogen oxidation as it relates to rotating reactor technology. In all cases, to achieve high degrees of nitrification, nitrifying organisms must be established and predominate on the media. Because heterotrophic organisms have substantially more rapid growth rates than nitrifying cultures, it is necessary to reduce the soluble carbonaceous BOD_5 to insure predominance of a nitrifying culture. This is achieved by design of substrate loading parameters so that, within the nitrifying reactor, soluble BOD_5 concentrations are equal to or less than 15 mg/L. At a concentration of 15 mg/L soluble BOD_5, the growth rate of the nitrifying microorganism can compete with the growth rate of the heterotrophic population. As the soluble BOD_5 is reduced below 15 mg/L, nitrifying organisms predominate, and ammonium oxidation proceeds at its optimum rate (Figure 5.18). The autotro-

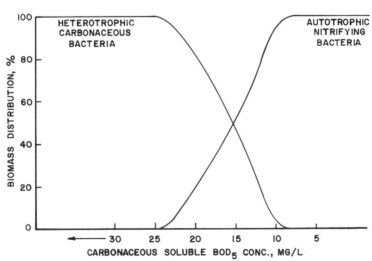

FIGURE 5.18. **Nitrifying population on a rotating biological reactor (domestic wastewater).**

phic nitrifying population present on the media are subject to environmental conditions that limit their growth. As a result, certain conditions such as pH, alkalinity, mixed liquor dissolved oxygen, and temperature must be considered during design to optimize the nitrifying growth rate.

Design of rotating biological reactor systems for nitrification. Through various full-scale, plant-scale, and pilot-plant experiences, it has been determined that ammonium removal rate proceeds as a zero order reaction to a concentration of 4 to 5 mg/L NH_3-N within the reactors and a first order or concentration-dependent reaction below 4 to 5 mg/L (Figure 5.19). Design of a rotating reactor is based on these kinetic rates. It is assumed there is no need for staging in reactors with NH_4-N concentrations exceeding 5 mg/L and that some degree of staging should occur below 5 mg/L NH_4-N.

Surface Area Determination. When designing for combined BOD_5 removal and nitrification, a two-step design procedure is necessary. Significant nitrification will not occur in the rotating biological process until the soluble BOD_5 concentration is reduced to 15 mg/L or less. Using general BOD_5 and NH_4^+-N curves (Figures 5.20 and 5.21), the first calculation is to determine the media area necessary to reduce the soluble BOD_5 concentration using Figure 5.20. For influent ammonium nitrogen concentrations of 15 mg/L and above, it is necessary to reduce the soluble BOD_5 concentration to the same value as the ammonium nitrogen concentration. The second calculation then uses a nitrification design curve, Figure 5.21, to determine the reactor media area necessary to reduce the influent ammonium nitrogen concentration to the required effluent concentration. The sum of the two surface area calculations then indicates the total surface area required for the combined BOD_5 removal and nitrification. If low-temperature conditions exist, surface area calculations must be adjusted. Typical temperature correction

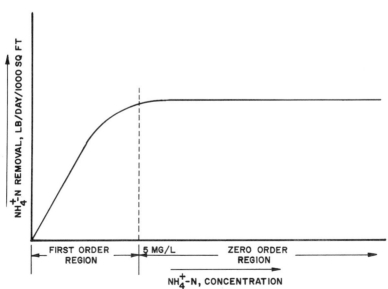

FIGURE 5.19. Single stage substrate removal, rotating biological contactor. NOTE: lb/day/1000 sq ft $\times$ 4.88 = g/m$^2 \cdot$d.

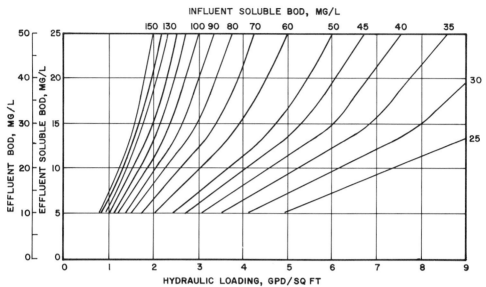

FIGURE 5.20. Domestic wastewater BOD removal with a rotating biological reactor (T > 55°F). NOTE: gpd/sq ft × 40.74 = L/m³·d.

factors are shown in Figures 5.22 and 5.23. Above 13°C, temperature correction is unnecessary.

For applications where high degrees of both BOD_5 removal and nitrification are required, the effluent ammonium nitrogen concentration invariably controls the overall rotating biological reactor design. For those cases where the required effluent total BOD_5 is less than 10 mg/L, tertiary filtration normally will be required to produce consistently

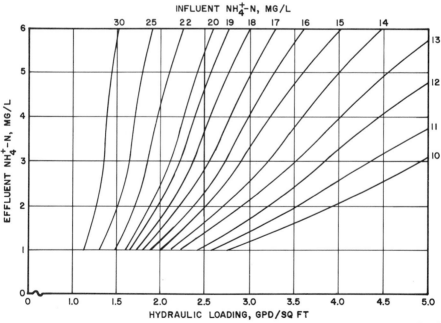

FIGURE 5.21. Nitrification of domestic wastewater using a rotating biological reactor (T > 55°F). NOTE: gpd/sq ft × 40.74 = L/m²·d.

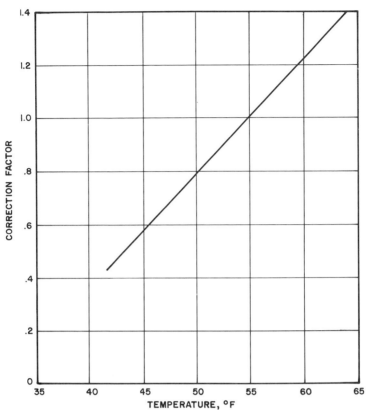

FIGURE 5.22. Temperature correction for nitrification using a rotating biological reactor. NOTE: 0.555 (°F − 32) = °C.

the required effluent. In this case, the overall hydraulic loading required to meet the ammonium nitrogen effluent level will very likely control media requirements. For applications requiring a high degree of BOD$_5$ removal, but only a moderate degree of nitrification, it will be necessary to separately calculate the overall surface area requirements to meet the BOD$_5$ removal and to meet ammonium nitrogen removal. The lower of the two required hydraulic loadings then will determine the surface area requirements.

pH. Optimum rates for nitrification have been determined during other investigations at pH 7.5 to 8.6. Full-scale determinations on rotating reactors indicate little effect on biological kinetics at pH ranges of 7.2 to 8.6. However, pH values below 7.2 tend to retard process efficiencies. A pH range of 7.2 to 8.6 maintained within the nitrification section of the reactor is indicated.

Dissolved Oxygen. Low dissolved oxygen concentrations (below 1.5 mg/L reactor DO) have been demonstrated to retard the nitrification efficiency of the RBC process. As a result, it is recommended that reactor DO be maintained continually at greater than 1.5 mg/L.

Alkalinity. Another factor that must be given consideration in nitrification design is wastewater alkalinity. Biological oxidation of ammonium nitrogen generates hydrogen ions that must be neutralized by the available alkalinity in the wastewater. Approximately 7 mg of carbonate alkalinity is required to neutralize the acidity generated from the oxidation of 1 mg of ammonia nitro-

57

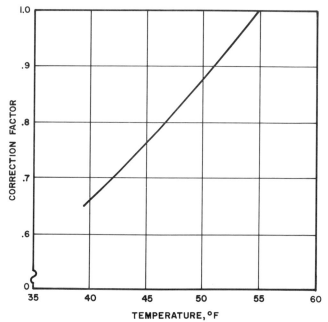

FIGURE 5.23. **Temperature correction for BOD removal. NOTE: 0.555 (°F − 32) = °C.**

gen. It is also recommended that a design residual alkalinity level of greater than 50 mg/L be maintained to avoid problems of fluctuations in plant influent alkalinity. If the wastewater contains insufficient alkalinity for the amount of ammonium nitrogen to be oxidized, the wastewater pH will be depressed and nitrification efficiency could be affected adversely. For applications where water supplies have low alkalinity or where there is extensive use of water softening, it may be necessary to supplement the wastewater alkalinity through chemical addition.

Load Variations. No adjustments to design hydraulic loading (Figure 5.21) are required as long as the peak : average flow ratio is 2.5 or less. When designing for nitrification, it is important that all potential sources of ammonium, particularly those from sludge-conditioning supernatants, are included in the influent ammonium nitrogen design conditions. This will be especially important when considering the use of anaerobic

sludge digestion or thermal sludge conditioning because the supernatants from these processes can contain several hundred mg/L of ammonium nitrogen. When using these processes for solids treatment it is important that the return ammonium load be equalized either in a separate equalization facility or that the liquors be returned during periods of low loading on the treatment plant in an effort to provide a relatively constant ammonium loading to the reactor. This will avoid any breakthrough of ammonium nitrogen in the final effluent.

Effects of Staging. Because ammonium nitrogen removal in the process above 5 mg/L is zero-order, that is, it proceeds at a constant rate, staging of the reactor media is not important when producing effluent concentrations of 5 mg/L or above. For effluent concentrations below 5 mg/L, staging of the media is important. To maximize the efficiency of the media area, it is recommended that multiple stages be used for nitrification of low ammonium concentrations.

High Ammonium Concentration—For wastewater containing more than 30 mg/L ammonium nitrogen, a two-step design procedure is required. At concentrations beyond the range of Figure 5.21 (30 mg/L), the process removes ammonium at a constant rate of 1.46 g/m²·d (0.3 lb/d/1000 sq ft) at 12.8°C. The first step in the calculation is to determine the surface area required to reduce the ammonium concentration from the inlet value of 30 mg/L at the same constant rate. The balance of removal will then be performed by additional surface area determined from Figure 5.21. The total surface requirement is the sum of the two surface area calculations. Any required temperature corrections are made to the total surface area requirement.

High ammonium concentrations often may be accompanied by high total Kjeldahl nitrogen concentrations. This is particularly true for combined domestic and industrial wastes and on wastewaters from small residential communities. In these applications it is imperative to determine the soluble organic forms and the hydrolyzable forms of suspended organic nitrogen. These two nitrogen forms are oxidizable by nitrifying cultures and should be added to the ammonium nitrogen when evaluating influent oxidizable nitrogen forms for design calculations.

Packed-bed reactors (PBR). Packed-bed or submerged filters, as they are sometimes referred to, are relatively new unit processes. One configuration of a packed-bed reactor is shown in Figure 5.24. Wastewater is introduced into the bottom of the reactor through a distribution system and flows upward through a bed of media that provides a surface area for biological growth. Hydraulic loadings normally fall within the range of 0.07 to 1.4 L/m²·s (0.1 to 2.0 gpm/sq ft). Media including 2.5- to 3.8-cm (1- to 1.5-in.) stones, 0.5-cm gravel, 1.8-mm (effective size) anthracite, and 9-cm "Maspac," have all been used successfully as media for PBRs. To date, these systems have been run only as separate stage processes. The frequency of backwash, which is required to reduce clogging and build-up of pressure losses, will vary, depending on influent quality and flow rate. With the plastic media, the frequency can be reduced considerably because of the high void volume. In

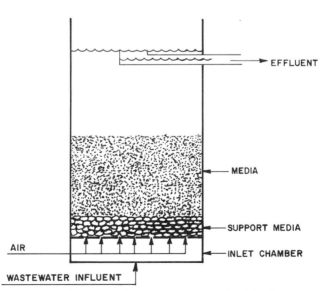

FIGURE 5.24. Schematic diagram of a packed-bed reactor (PBR).

most cases excess solids in the higher void space media (80 to 95%) can be withdrawn simply by draining the units on a monthly or less frequent schedule. To avoid short-circuiting, good hydraulic distribution of the influent is necessary along with the ability to backflush the media when plugging occurs.

With gravel media, standard practice is to backwash the reactor at approximately 17 L/m^2·s (25 gpm/sq ft) at least three times per week and, in some cases, daily.

Data available for formulation of design criteria for PBR units are summarized in Table 5.5. Loading rates fall in the range of 0.06 to 0.43 kg/m^3·d of NH$_4^+$-N (4 to 27 lb/d/1000 cu ft) oxidized. Factors affecting the oxidation rate are the influent quality (BOD$_5$, TKN, and NH$_4^+$-N), temperature, and the type and surface area of the media selected as a biological-growth surface.

Methods of Supplying Oxygen. Several techniques have been used to supply the necessary oxygen required for nitrification. Air of high purity oxygen may be bubbled directly into the reactor, or the wastewater may be preoxygenated with high purity oxygen prior to entry into the reactor. The preoxygenation techniques can supply only enough oxygen to oxidize between 5 and 8 mg/L of ammonium. Because of this, effluent recirculation of 2:1 to 4:1 is usually required when preoxygenation is used. Bubbling air or oxygen offers the advantage of preventing or reducing solids build-up in the filter, reducing backwashing requirements. However, the suspended solids in the effluent are usually higher than if preoxygenation is used.

Biological fluidized bed reactors. The most recent entry to the commercial market is the biological fluidized bed process. To date only pilot work has been reported for nitrification of municipal and industrial wastes. One full-scale

plant, however, has been constructed at the Dworshak National Fish Hatchery near Lewiston, Idaho, for the treatment of 1300 L/s (30 mgd) of fish hatchery reuse water. The system is designed to oxidize an inlet ammonia concentration of 0.5 mg/L by 60 to 80%. The 10 to 15°C water is evenly distributed to 2.1- to 4.2-m (7 to 14 ft) diameter reactors, five of which are 8.5 m (28 ft) high while two are 7.3 m (24 ft) high. Effluent from the reactors is sent to an indoor nursery system and a larger outdoor system of rearing ponds.

Briefly, the process consists of passing the wastewater to be treated upward through a bed of fine grained media, such as sand, at a velocity sufficient to impart motion to or fluidize the bed. This allows the entire surface area of each sand grain to be in contact with the wastewater and available for biological growth. Specific surface areas of the media range from 2000 to 3300 m^2/m^3 (600 to 1000 sq ft/cu ft). Because of the extremely high surface area and resultant biomass concentrations (8000 to 20 000 mg/L of VSS), treatment times and required reactor sizes are greatly reduced. Because the media is in a continual state of motion with upflow rates between 4.7 and 16.9 L/m^2·s (7 and 25 gpm/sq ft), plugging or channeling is minimized. As the growth on the particles exceeds that necessary to maintain process efficiency, the particles are removed from the reactor, the excess growth sheared and separated from the sand grains, and the partially "cleaned" sand grains returned to the reactor. The sludge is disposed of in a concentrated stream of less than 0.1% of the average flow when treating secondary effluent. No clarification is required following this process. A schematic flow chart of the process is shown in Figure 5.25.

All pilot work to date has used preoxygenation of the wastewater prior to entry into the reactor. Recirculation of effluent is used to satisfy the oxygen demand and nearly total nitrification can be attained.

TABLE 5.5. Packed-bed reactor performance when treating secondary effluents.

Location and type of aeration	Media depth, ft (m)	Media type	Surface loading gpm/sq ft (m³/m²/d)	Empty bed hydraulic detention time, min	Ammonia-N oxidation rate lb/day 1000 cu ft (kg/m³/d)	Temp., °C	Influent Quality, mg/L				Effluent Quality, mg/L						Removals, percent			
							BOD.	SS	Organic-N	Ammonia-N	BOD.	SS	Organic-N	Ammonia-N	Nitrite-N	Nitrate-N	BOD.	SS	Organic-N	NH₃-N
Union City, Calif. preoxygenation (oxygen)	3.0 (.91)	1 to 1.5 in. (2.5 to 3.8 cm) stone	.15 (8.8)	154	7.7 (.12)	21 to 27	35	27	3.6	14.3	5	4	1.5	1.0	0.6	15.9	86	87	58	93
			.21 (12)	103	12.1 (.19)	21 to 27	38	38	—	19.6	3	7	—	5.6	4.1	6.9	91	83	—	71
			.29 (17)	77	9.8 (.15)	21 to 27	37	25	5.7	15.2	10	7	2.5	6.9	1.7	6.0	74	74	56	55
Bubble (oxygen)	3.0 (.91)	1 to 1.5 in. (2.5 to 3.8 cm) stone	.15 (8.8)	154	9.7 (.15)	16 to 30	37	30	5	18.3	10	16	2.2–4.7	1.8	0.4	18.3	74	48	—	91
			.29 (17)	77	—	16 to 30	43	48	—	—	25	51	—	—	—	—	41	−6	—	—
Ames, Iowa bubble (air)	5.0 (1.5)	1.8 mm	1.0 (59)	37	9.4 (.15)	21 to 23	39	43	—	8.4	19	—	—	5	—	—	51	20	—	46
		anthracite	0.4 (23)	94	5.9 (.09)		20	26	—	6.8	5	—	—	1	—	—	77	32	—	90
	8.0 (2.4)	Maspac[a]	1.0 (59)	60	5.1 (.08)	21 to 23	39	43	—	8.4	19	—	—	5	—	—	51	42	—	40
			0.4 (23)	150	3.5 (.06)		20	26	—	6.8	8	—	—	1	—	—	59	40	—	87
	13 (4.0)	Series anthracite	0.5 (29)	195	6.1 (.10)	9 to 14	26	47	—	14.4	8	—	—	1	—	—	68	49	—	92
		Maspac[a]	0.75[b] (44)	130	4.6 (.07)	12	37	63	—	11.2	16	—	—	5	—	—	56	39	—	59
Pomona, Calif. bubble (oxygen or air)	5.5 (1.7)	5 mm gravel	0.75 (44)	55	26.5 (0.42)	27 to 28	7[b]	9[b]	—	18.1	—	—	—	1.9	0.6	16.6	—	—	—	90
			0.59 (35)	70	20.1 (0.32)	25 to 26	—	—	—	17.6	—	—	—	1.4	0.6	16.3	—	—	—	92
			0.41 (24)	100	13.3 (0.21)	19 to 22	—	—	—	16.8	—	—	—	2.0	0.9	16.9	—	—	—	88
			0.46 (27)	90	14.7 (0.24)	20 to 25	—	—	—	16.4	—	—	—	1.7	0.4	15.6	—	—	—	90
			0.39 (23)	105	16.4 (0.26)	20 to 22	—	—	—	20.7	—	—	—	1.5	0.5	20.7	—	—	—	93
	11.0 (3.4)	5 mm. gravel. two columns in series	1.49 (87)	55	26.7 (0.43)	26 to 28	—	—	—	17.6	—	—	—	1.3	0.4	17.1	—	—	—	93
			0.75 (44)	110	13.8 (0.22)	22 to 25	—	—	—	18.9	—	—	—	1.9	0.2	18.2	—	—	—	90

[a] A product of the Dow Chemical Co., Midland, Mich.
[b] Average for test series treating activated sludge effluent.

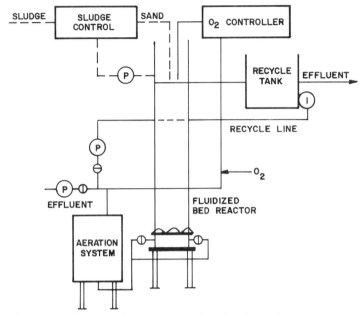

FIGURE 5.25. Schematic diagram of a fluidized bed reactor (FBR).

Combined carbon oxidation-nitrification. Several pilot studies have been run using the biological fluidized bed in the combined mode. Pertinent operating conditions and results typical of one study are listed in Table 5.6. Generally, greater than 80% BOD$_5$ and ammonium removals were achieved without any secondary clarification. BODs were reduced from 128 to 23 mg/L and ammonium from 25.6 to 5.3 mg/L. At the time this manual went to press, insufficient data were available to provide a design procedure and define areas of best applications and limitations.

Separate stage nitrification. Several studies have been reported using the fluidized bed for separate stage nitrification. A design removal curve for the nitrification of secondary effluent is presented in Figure 5.26.

ECONOMICS

Biological nitrification versus physical chemical nitrification. Cost is often the single most influential factor in selection of a nitrification process. Biological nitrification is generally recognized as the least costly form of ammonium removal. Lime precipitation–air stripping, breakpoint chlorination, and ion

TABLE 5.6. Combined BOD/NH$_4^+$ removal.

Parameter	Operating values
Inflow (gpm)	3.1
Recycle (gpm)	21.0
Detention time (min)	72
Temperature (°C)	12.5
Influent BOD$_5$ (mg/L)	128
Effluent BOD$_5$ (mg/L)	23
Filtered effluent BOD$_5$ (mg/L)	4
Influent TSS (mg/L)	90
Effluent TSS (mg/L)	28
Influent NH$_4^+$-N (mg/L)	25.6
Effluent NH$_4^+$-N (mg/L)	5.3
Influent TKN (mg/L)	37.6
Effluent TKN (mg/L)	10.0
Influent NO$_2$ + NO$_3$ (mg/L)	0
Effluent NO$_2$ + NO$_3$ (mg/L)	18.6
SRT (days)	15.4
F/M	0.12
MLVSS (mg/L)	20 000

NOTE: gpm $\times$ 6.308 $\times$ 10^{-2} = L/s.

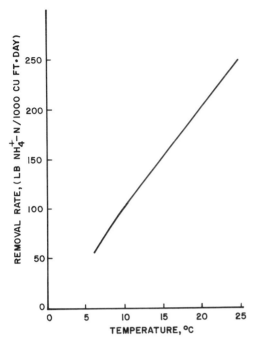

FIGURE 5.26. Nitrification removal versus temperature for a fluidized bed reactor.

exchange are more costly than biological nitrification.

Year-round, low-temperature operation (less than 10°C) generally favors physical–chemical processes over biological nitrification because of the low nitrification rates. An exception is the air-stripping process, which is adversely affected by low temperature.

The presence of compounds toxic to nitrifiers in the influent may dictate against the choice of biological nitrification. A good industrial pretreatment program will exclude the majority of toxic substances from the collection system and treatment facility.

Combined versus separate stage nitrification. Suspended growth nitrification usually yields effluent of lower ammonium concentrations than effluent of attached growth processes. Two stage nitrification offers good protection against most toxicants, stable operation, and very low effluent ammonium

levels. The disadvantages of the system are that the sludge inventory requires careful control when the BOD_5/TKN ratio is low, and the stability of the operation is linked to effective operation of secondary clarifiers for biomass return. Also, separate stage operation usually is more costly than combined treatment.

Factors favoring combined carbon oxidation nitrification are:

● Use of biological solids retention time offers a more positive means of process control,

● Lower sludge quantities are wasted in combined carbon oxidation-nitrification systems than in separate stage treatment,

● Sludge settling generally is improved because of long solids retention times, and

● Process control is easier as a result of a lesser number of settling units and a lack of shifting sludge from one stage to another to maintain sludge inventory.

Factors favoring separate nitrification are:

● For reasons of economy, separate stage nitrification generally is preferred when industrial waste represents greater than 10% of the waste,

● Additional settling tanks afford flexibility of operation, and

● Presence of biological carbon oxidation units with intermediate clarification ahead of the nitrification units protects against toxicants and dampens peak load factors.

REFERENCES

1. "Wastewater Treatment Plant Design." Joint ASCE-WPCF Manual of Practice, MOP 8, Water Pollut. Control Fed., Washington, D.C. (1977).
2. "Process Design Manual for Nitrogen Control." U.S. EPA, Washington, D.C. (Oct. 1975).
3. Painter, "Process Design Manual for Nitrogen Control." U.S. EPA, Washington, D.C. (Oct. 1975).
4. Anthonisen, "Process Design Manual for Nitrogen Control." U.S. EPA, Washington, D.C. (Oct. 1975).

5. "Operation of Wastewater Treatment Plants." Manual of Practice, MOP 11, Water Pollut. Control Fed., Washington, D.C. (1976).

6. "International Seminar on Control of Nutrients in Municipal Wastewater Effluents." Vol. L, LL, and LLL. Proc., U.S. EPA, MERL, Cincinnati, Ohio (Sept. 1980).

7. "Converting Rock Trickling Filters to Plastic Media." U.S. EPA, Publ. No. EPA 600/2-80-120 (Aug. 1980).

8. Lawrence, A. W., and McCarty, P. L., "Unified Basis for Biological Treatment Plant Design and Operation. "Jour. San. Eng. Div., Proc. Amer. Soc. Civil Engr., **96,** 757 (1970).

Chapter 6

Nitrogen Removal

INTRODUCTION

Biological denitrification can be defined as the process by which microorganisms reduce nitrate ion to nitrogen (N_2), nitrous oxide (N_2O), or nitric oxide (NO).

In the 1940s, it was recognized that denitrification occurred in many activated sludge wastewater treatment plants.[1,2] This sparked considerable research in biological nitrogen removal. As a result, denitrification became the most widely used nitrogen removal process in municipal wastewater treatment. It is currently also the process of choice for many industrial wastes requiring nitrogen treatment.

A number of bacterial species that naturally occur in the activated sludge process are capable of denitrification. Included in the list of denitrifying bacteria identified to date are: Achromobacter, Bacillus, Brevibacterium, Enterobacter, Lactobacillus, Micrococcus, Paracalobactrum, Pseudomonas, and Spirillum.[3-5]

These organisms are termed heterotrophic bacteria because they can metabolize complex organic substrates. These bacteria are capable of using either molecular oxygen or nitrate oxygen as a terminal electron acceptor when they oxidize organic compounds. Under anoxic conditions (such as in the absence of free molecular oxygen), denitrifying bacteria reduce nitrate by a process called nitrate dissimilation in which nitrate or nitrite replaces oxygen in cell respiration. The process of nitrate dissimilation occurs through a complex series of reactions catalyzed by enzymes. These processes, although not fully understood, are dealt with at length in the literature[6-8] concerning nitrogen metabolism.

Simply stated, nitrate dissimilation takes place in two steps. First, nitrate is reduced to nitrite through the transfer of two electrons from the organic substrate[9] producing energy for cell synthesis. All nitrate-reducing bacteria produce nitrite as the first reaction product. In a second step, nitrite is reduced to nitric oxide (NO), nitrous oxide (N_2O), or N_2, nitrogen gas, being the predominant products.

Denitrifying bacteria can also convert nitrite by the process known as assimilation to ammonium, NH_4^+-N, for cell synthesis. This does not normally occur if the ammonium present in the wastewater is sufficient to meet growth requirements.

The biochemical processes used for electron transfer from organic carbon to oxygen are very close to nitrate dissimilation. In fact, the same series of enzymatic reactions occur in both processes. The key difference between oxygen respiration and denitrification results from a single enzyme, nitrate reductase, produced in the absence of oxygen that completes the electron transport process required for nitrate dissimilation.

The similarity of the two processes accounts for the fact that facultative bacteria in a single sludge process are able to switch easily from oxygen to nitrate as electron acceptors in the oxidation of organic carbon compounds.

Effect of dissolved oxygen. Oxygen suppresses production of a critical enzyme in the electron transport system required for denitrification. Use of oxygen as electron acceptor also yields more free energy[10] than nitrate. Therefore, oxygen respiration is favored when both are present.

A number of researchers, however, have observed denitrification at positive dissolved oxygen concentrations.[11-14] The occurrence of denitrification under apparent aerobic conditions is explained by the existence of an oxygen gradient within the bacterial floc that results in low dissolved oxygen concentration at the floc center.[15,16] This phenomenon is worthy of mention because in some cases it will result in enhanced nitrogen removal efficiency in the nitrification-denitrification process. The oxidation ditch process modified for nitrification–denitrification and single sludge pre-denitrification flow sheets have exhibited significant denitrification levels in the aerobic portion of the process.[17,18]

Alkalinity and pH. Denitrification is a net producer of alkalinity. Carbonic acid is converted to bicarbonate[19] as a result of denitrification of nitrate to nitrogen gas. The alkalinity production determined by different researchers[20,21] is on the order of 2.9 to 3.0 mg of $CaCO_3$ per mg of nitrogen reduced. The production of alkalinity will raise system pH and offset some of the loss of alkalinity from nitrification in combined systems.

The optimum pH for denitrification has been determined in a number of studies[22-24] to lie between 7 and 8. The specific optimum will vary depending on bacteria present and wastewater components. Because nitrifying treatment plants operate close to optimum pH for denitrification, no attempt is made to control this variable. This includes treatment plants with iron or aluminum salt additions. The only exception might be a system in which chemical phosphorus precipitation with lime at high pH is being combined with nitrogen removal.

Temperature. Denitrification has been reported in the literature over a range of temperatures from 0° to 50°C.[25] Temperature affects both the microbial growth rate and the removal rate of nitrate. Both factors are taken into account in designing denitrification systems. Over the normal temperature span for wastewater treatment, from 5° to 30°C, denitrification follows an Arrhenius temperature relationship[26] commonly expressed in the following form:

$$K = Ae^{-E/RT} \qquad (1)$$

where

K = the reaction velocity, mol/s,
A = a constant,
E = the activation energy, cal/mol,
R = the gas constant, 0.082 06, L·atm/mol·K, and
T = the absolute temperature, K.

A modified Arrhenius expression[27] used to express the temperature dependence of denitrification is as follows:

$$K_T = K_{20}\theta^{T-20} \qquad (2)$$

where

K_T = the reaction velocity at temperature T°C, day^{-1},
K_{20} = the reaction velocity at 20°C, day^{-1}, and
θ = temperature coefficient, varying from 1.025 to 1.15.

Table 6.1 shows the activation energy and temperature coefficients determined by a number of researchers over temperatures from 3° to 30°C.[26,28-30] These data point out the sensitivity of denitrification even at the low temperatures encountered under winter conditions.

Effect of carbon and nitrate concentration on kinetics. The combined kinetic expression used to describe rate of denitrifier growth, and, therefore, nitrate removal, is shown as follows:

$$\hat{u}_{DN^*} = u_{DN}\frac{N}{K_N + N} \cdot \frac{C}{K_C + C} \qquad (3)$$

TABLE 6.1. Temperature coefficients and denitrification.

Reference		Temperature range, °C	E, cal/mole	θ
Dawson[28]	Batch	3–27	16 800	1.12
	P. denitrificans	10–20		1.10
Stensel[29]	Batch activated sludge SRT = 2 days	15–25	10 000	1.06
	Continuous activated sludge	10–20	19 500	1.13
Mulbarger et al.[30]	Activated sludge SRT = 7.6 days	10–20	19 000	1.15
Sutton et al.[26]	Activated sludge, batch and continuous, 3-day sludge age	6–25	15 300	1.09
	Activated sludge, batch and continuous 6-day sludge age	6–25	15 900	1.10

where

$\hat{u}_{DN^*}$ = maximum growth rate of denitrifiers for a given temperature and pH,

u_{DN} = actual denitrifier growth rate determined by temperature pH, nitrate and carbon concentrations,

N = nitrate concentration mg/L,

K_N = half saturation constant for nitrate, mg/L,

C = organic substrate concentration, mg/L, and

K_C = half saturation constant for organic substrate, mg/L.

Studies carried out on denitrifying systems using methanol have shown that a very small excess of organic carbon will ensure denitrification rates close to maximum. The half saturation constant for methanol was determined to be 0.1 mg/L.[31] This low value means that an excess of 1 mg/L of methanol or other organic substrates in anoxic reactor effluent will give denitrification rates at more than 90% of the maximum rate.

The same has been found of nitrate concentration. The half saturation constant has been determined to be 0.16 mg/L NO_3-N with solids recycle at 20°C for suspended growth systems.[34,35] For attached growth, K_N was found to be 0.06 mg/L NO_3-N.[36,37] Because K_N is so low, nitrate concentration has very little effect on rate of reaction above 1 mg/L NO_3-N and for design purposes can be considered a zero-order reaction. This phenomenon accounts for the low nitrate levels attainable in the effluent from denitrifying systems.

FIXED-FILM REACTORS

Denitrification with attached-growth systems. There are a variety of attached-growth systems currently available to denitrify domestic and industrial wastewaters. These include biological fluidized beds, packed (fixed) beds of either highly porous media or low porosity, fine media, nitrogen gas-filled packed beds,

and rotating biological contactors. Design for these reactors is based on a combination of theoretical considerations and empirical data. Design loadings normally are expressed in terms of mass of NO_3-N removed per volume of reactor or surface area available for biological growth.

As is the case for suspended growth systems, care must be taken to exclude oxygen from the reactor. Oxygen, if available, will be preferentially used in place of nitrate and nitrite, thereby increasing methanol requirements and sludge production. Excluding oxygen is normally accomplished by completely submerging the reactor media in the wastewater, although one process does trap the nitrogen gas produced to form an oxygen-free atmosphere within the reactor.[36,37]

High-porosity—packed beds. High-porosity systems have been tested in pilot-[38-42] and full-scale studies.[43] A schematic of one possible mode of operation is shown in Figure 6.1.[31] Media may be either bundled corrogated sheet or random dump plastic typically associated with trickling filters. To obtain sufficient contact time and surface area, a series configuration of two or more reactors is sometimes used.

Because a high void ratio (90 to 95%) is provided and maintained, plugging is minimized. The biomass is either allowed to slough, as in a trickling filter operation, or is removed by periodic backwashes. In either case, backwashing, although it may be infrequent, is still required. At El Lago, Tex., where flexible-ring media were used, backwashing was performed routinely every 4 weeks. A combined water and air backwash was used. Water and air backwash rates were 6.8 L/m²·s (10 gpm/sq ft) and 3.6 m³/m²·min (10 cfs/sq ft), respectively. Backwashing was primarily performed to prevent high effluent suspended solids caused by sloughed growth.[43] Surface removal rates observed from various sites are summarized in Figure 6.2.[31]

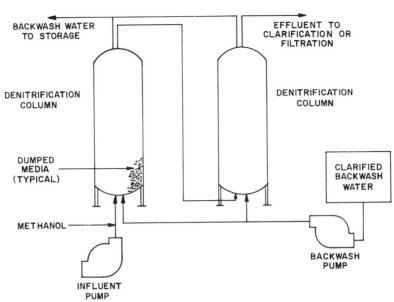

FIGURE 6.1. Typical process schematic for submerged high-porosity media columns.

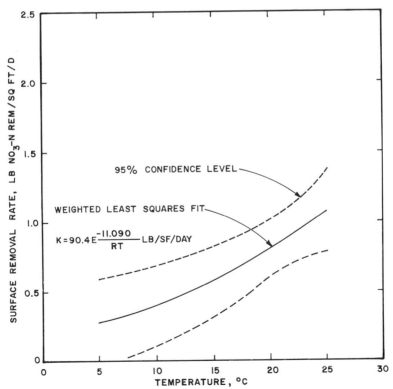

FIGURE 6.2. Surface denitrification rate for submerged high-porosity media columns, Hamilton Ontario. Specific Surface is 4700 to 9000 m^2/m^3 (142 to 274 sq ft/cu ft); voids is 70 to 78%. NOTE: lb/day/sq ft $\times$ 4.88 $=$ kg/$m^2\cdot$d.

Design of a system should be based on the minimum temperature and peak nitrate loading. From this a required surface area may be obtained (Figure 6.2). The reactor volume then can be determined by choosing a media of known specific surface.

Low-porosity/fine-media packed beds. Media used in these systems normally consist of fine-grained gravel or sand, usually between 2 and 5 mm in diameter. This type of system, in addition to denitrifying the wastewater, provides filtration. This, however, has its drawbacks, as the reactors must be backwashed frequently to prevent excessive headloss from occurring. The frequency is dependent on the nitrate and suspended solids loadings applied to the system, as well as the media size used.

One design consists of 1.8 m (6 ft) of uniformly graded sand 2 to 4 mm in size. Hydraulic application rates for removing 20 mg/L of NO_3-N from municipal wastewater range from 0.7 to 1.7 $L/m^2 \cdot s$ (1.0 to 2.5 gpm/sq ft) for minimum wastewater temperatures of 10° and 21°C, respectively.[44] A combined air and water backwash is recommended. Suggested backwash rates are 1.8 $m^3/m^2 \cdot min$ (6 cfm/sq ft) and 5.4 $L/m^2 \cdot s$ (8 gpm/sq ft) for air and water, respectively.

Nitrogen gas has been found to build up in the reactor between backwashes. This can cause short-circuiting. A "bumping" or short backwash is frequently used to free this trapped gas. Backwash rates of 5.4 to 10.8 $L/m^2 \cdot s$ (8 to 16 gpm/sq ft) for 1 or 2 minutes are required every 4 to 12 hours.[31] Backwash frequency will vary with the applied solids and nitrate loading rates. Temporary, partial inhibition of denitrification has been observed when combined air and water backwash was used. No inhibition has been detected when only water backwash is used.

One company has tested various combinations of media and determined the two listed in Table 6.2 as providing best overall performance. Hydraulic applica-

TABLE 6.2. Media designs for denitrification.

Filter material	Layer depths, in. (cm), of two media combinations tested	
	F-II	F-II
Garnet sand; $d_{10} = 0.27$ mm	3 (7.6)	3 (7.6)
Silica Sand; $d_{10} = 0.5$ mm	9 (22.9)	9 (22.9)
Anthracite coal; $d_{10} = 1.05$	18 (45.1)	8 (20.3)
Anthracite coal; $d_{10} = 1.75$	—	16 (40.6)

d_{10} = effective size.

tion rates reported by this manufacturer are in the same range as rates earlier reported, which are from 1 to 2 $L/m^2 \cdot s$ (1.5 to 3.0 gpm/sq ft).[31]

With effective backwashing, suspended solids removal efficiency is very good for both systems. Effluent suspended solids normally are under 10 mg/L with a reported range of 1 to 17 mg/L.[31]

Biological fluidized bed. In this process, the wastewater to be treated is passed upward through a bed of fine-grain material, such as sand, at sufficient velocity to "fluidize" or impart motion to the media. This frees the entire surface area of the media for biological growth. Because the system is fluidized, there is no danger of clogging. Therefore, extremely small media sizes can be used. This significantly increases the specific surface and allows extremely high biomass concentrations to be maintained within the reactor. Biomass concentrations between 10 000 and 40 000 mg/L have been reported.[44-46] This type of system never requires backwashing as it is operating in a continuous backwash condition. In one study at Nassau

County, N.Y., hydraulic flux rates of 10.2 and 16.3 L/m^2·s (15 and 24 gpm/sq ft) were successful in achieving virtually complete denitrification of municipal wastewater in empty bed detention times of approximately 6.5 and 4.0 minutes, respectively.[45] Particles laden with excess biological growth are removed periodically from the system by a pump, mechanically and hydraulically abraded to remove the excess biomass, separated from the waste sludge, and returned to the reactor. Because of this, no clarification is necessary to remove excess solids. Influent and effluent solids concentrations have been shown to be virtually identical.[46]

At Oak Ridge National Laboratories, the fluidized bed has been successfully used to treat high-strength wastewaters with nitrate concentrations of over 2000 mg/L NO$_3$-N.[47]

Figure 6.3 is a design loading curve for denitrification of municipal wastewater, showing the effect of temperature on design loading rate.[45]

A schematic of the pilot system used in one study is shown in Figure 6.4.[45] This same basic flow scheme was used in the design of a 91 mL/d (24-mgd) municipal plant in Pensacola, Fla.[48] The Pensacola plant incorporates chemical treatment for 5-day biochemical oxygen demand (BOD$_5$) and suspended solids removal and a pure oxygen suspended growth system for nitrification. The denitrification is accomplished in four 5.8-by-5.8-m (19-by-19-ft) deep reactors. Designed at an average flux rate of 10.2 L/m^2·s (15 gpm/sq ft), the reactors use

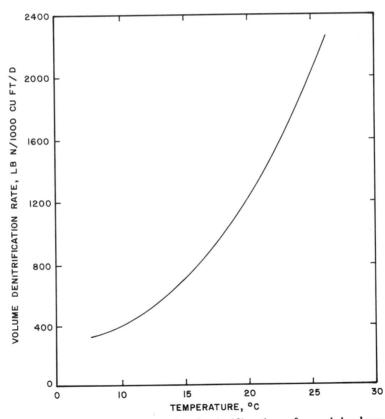

FIGURE 6.3. Design loading curve for denitrification of municipal wastewater in a fluidized bed reactor. NOTE: lb/day/1000 cu ft × 16 = g/m^3·d.

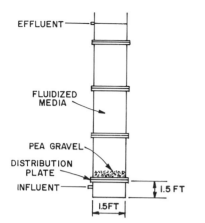

EFFLUENT

FLUIDIZED MEDIA

PEA GRAVEL

DISTRIBUTION PLATE

INFLUENT

1.5 FT

1.5 FT

FIGURE 6.4. Schematic of fluidized bed reactor. NOTE: ft × 0.30 = m.

sand as the fluid bed medium. It was estimated that a 90% reduction in land requirements was realized over the comparable suspended growth system. Because of its small size and the requirement for pilings at this site, the great reduction in area resulting from the use of the fluidized bed system saved the community about $1.5 million in piling and site preparation costs. However, the use of methanol to denitrify the wastewater was a significant cost factor and must be evaluated versus the alternative processes. A summary of biological fluidized bed denitrification studies at several locations is shown in Table 6.3.

Nitrogen gas filled—packed bed. In this system, the media is not submerged as in all the other columnar reactors. Instead, the reactor is operated as a nitrogen-gas-filled trickling filter. The nitrogen gas generated is trapped within the reactor. In this type of system (Figure 6.5), influent wastewater is spread over the top of the media and flows as a thin laminar film over the biologically coated media. Sufficient void space is maintained to prevent plugging or ponding resulting from biological growth. Sloughed solids must be removed in a clarifier following the reactor.

Much pilot testing has been done to establish optimum surface area of the media and loading rates.[36,37] Initial testing was done with a media with a specific surface of 223 m^2/m^3 (68 sq ft/cu ft). At high application rates this media clogs. It was replaced with media that had a specific surface of 138 m^2/m^3 (42 sq ft/cu ft) that functioned without problems. A loading rate of 19 250 $L/m^3 \cdot d$ (144 gal/cu ft·day) has been suggested as design criterion that will achieve desired removal efficiencies and permit the reactor to function without excess build up of growth on the media. However, pilot plant studies are encouraged to assist in the final sizing criteria selection at the time of preparing this manual (1983).

Rotating biological contactors. Besides the various columnar configurations, RBCs may also be used for denitrification. Operation of the contactors is similar to that for aerobic processes, except the media is totally submerged to avoid oxygenating the liquid. Antonie[48] has performed pilot work and presented a design loading curve (Figure 6.6). Clarification is required following the biological treatment stage to remove sloughed excess biomass. Pilot studies are recommended to define process kinetics.

SUSPENDED GROWTH REACTORS

Design methodologies. Design of suspended growth biological denitrification systems is normally based on either SRT (solids retention time or sludge age) considerations or the unit denitrification rate approach. The SRT approach is based on the growth rate of the denitrifying organisms, while the unit removal rate approach is based on the total biomass inventory. The unit denitrification rate is expressed as the mass of equivalent nitrate nitrogen removed per day per mass of total volatile suspended solids. The term "equivalent nitrate-nitrogen" accounts for the presence of other electron acceptors such as molecular oxygen or nitrite.

TABLE 6.3. Summary of biological fluidized bed denitrification.

Location	Reactor Surface Area (m²)	Bed Depth (m)	Media Type	Media Size (mm)	Type of Waste	Temp. (°C)	Carbon Source	HRT (min.)
Low strength (Domestic—River Water)								
Nassau Co.	0.21	3.6	sand	0.8	Domestic	20	MeOH	6.5
Nassau Co.	0.21	3.6	sand	0.8	Domestic	12.5	MeOH	6.5
Stevenage, England	0.03	3.1	sand	—	Domestic	16.4	sewage	3.9
Stevenage, England	0.03	3.1	sand	—	Domestic	14.4	sewage	5.5
Stevenage, England	0.03	3.1	sand	—	Domestic	11.2	sewage	3.2
Thames R., Eng.	0.03	3.6	sand	0.2–0.5	River Water	10	MeOH	11
Thames R., Eng.	0.03	3.6	sand	0.2–0.5	River Water	20	MeOH	11
High strength industrial wastes								
Oak Ridge	0.01	6.0	coal	0.25–0.6	Simulated Nuclear Waste	22	MeOH	8.5
Oak Ridge	0.01	6.0	coal	0.25–0.6	Explosives Wastewater	22	MeOH	8.5
Johannesburg, S.A.	0.13	6.0	sand	1.0	Wastewater	25	Molasses	6.0
						38	Molasses	6.0

Location	Flux rate (m/min)	MLVSS (mg/L)	DN rate (gN/m³-hr)	DN rate (grN/grVSS-day)	Nitrate Nitrite Removal (%)	NO₂⁻ + NO₃⁻-N (mg/L) Inf	Eff
Low strength (Domestic—River Water)							
Nassau Co.	0.6	30 000	226	0.18	98	21.5	0.2
Nassau Co.	0.6	30 000	190	0.15	98	18.3	0.4
Stevenage, Eng.	0.8	15 200	201	0.34	77	15.5	3.6
Stevenage, Eng.	0.55	27 400	85	0.07	95	8.2	0.4
Stevenage, Eng.	0.96	21 800	132	0.16	64	10.0	3.6
Thames R., Eng.	0.34	12–15 000	160	0.2	90	—	—
Thames R., Eng.	0.34	12–15 000	310	0.4	90	—	—
High strength industrial wastes							
Oak Ridge	0.51	—	2 100	—	88	215	42
Oak Ridge	0.51	—	1 110	—	100	215	0
Johannesburg, S.A.	1.3	—	533	—	90	—	—
	1.3	—	1 630	—	90	—	—

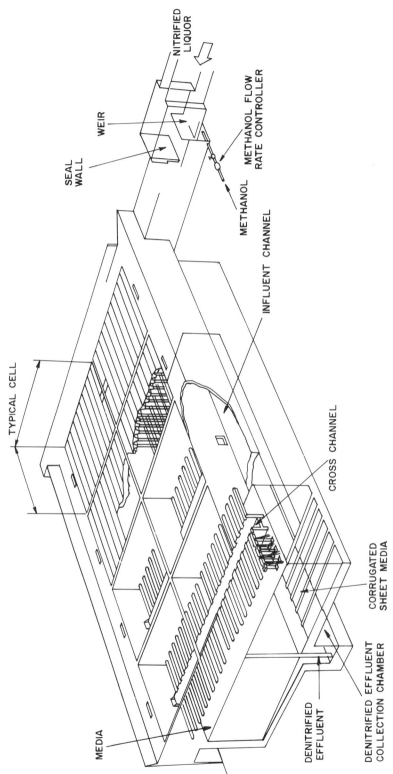

FIGURE 6.5. Nitrogen gas filled denitrification column.

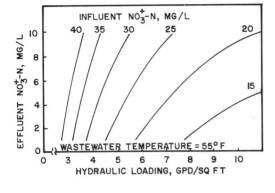

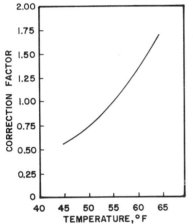

FIGURE 6.6. Design curves/denitrification, with temperature corrections. NOTE: gpd/sq ft × 40.74 = L/m²·d.

ulation occurs and the equation describing growth becomes:

$$\mu_{DN} = \hat{\mu}_{DN} \frac{N}{K_N + N} \cdot \frac{C}{K_C + C} \qquad (5)$$

where

μ_{DN} = growth rate of denitrifiers d^{-1},

$\hat{\mu}_{DN}$ = maximum growth rate of denitrifiers d^{-1},

N = nitrate-nitrogen concentration (mg/L),

K_N = half saturation constant equal to N concentration at which $\mu_{DN} = 0.5\ \mu_{DN}$ (mg/L),

C = carbon concentration (mg/L), and

K_C = half saturation constant (mg/L).

The value of K_C has been determined to be very low, approximately 0.1 mg/L for methanol, with similarly low values for other organics. This implies that it is possible to achieve near maximum denitrification rates with very little excess of carbon above the stoichiometric requirements. Under non-carbon-limiting conditions, Equation 5 reduces to:

$$\mu_{DN} = \hat{\mu}_{DN} \frac{N}{K_N + N} \qquad (6)$$

which is essentially a zero-order reaction.

Design of systems based on the SRT approach is normally accomplished using specific growth rate data from the literature or employing minimum solids retention time generated from bench-scale testing. To ensure consistent denitrification performance, a design safety factor should be incorporated. The safety factor is defined as:[31]

$$SF = \frac{SRT_{design}}{SRT_{min}} \qquad (7)$$

Application of a safety factor will ensure consistent denitrification at varying flows and nitrate concentrations, and the higher the SF, the lower will be the effluent nitrate level. From a practical viewpoint, a design SF of 2 is normally adequate. This should ensure an effluent NO₃-N of less than 0.5 mg/L.

In a denitrifying biological system operating under equilibrium conditions, the SRT is defined as

$$SRT = \frac{1}{\mu_{DN}} \qquad (4)$$

where

μ_{DN} is the specific growth rate of denitrifying organisms. From Equation 4, a design based on SRT control will establish the growth rate of denitrifying organisms.

The growth rate of denitrifying organisms is normally expressed via the Michealis-Menton relationship. Under most circumstances little or no nitrite accum-

There is a limited amount of data in the literature regarding design of denitrification systems based on SRT control. For a separate sludge system, SRT is defined as the mass of biological solids in the reactor divided by the total quantity of biomass lost or wasted per day from the system. However, for single sludge systems the design SRT is defined as:

$$SRT_{design} = \frac{V_{DN} \times SRT_s}{V_s} \qquad (8)$$

where

V_{DN} = volume of denitrification reactor,
V_s = total volume of the biological system, and
SRT_s = total system SRT, designed for minimum operating temperature.

In the treatment of municipal wastewater, Sutton[49] determined the minimum SRT necessary to achieve denitrification using the single sludge combined nitrification-denitrification flowsheet. Systems were operated in both the pre- and post-denitrification mode. Carbon sources were methanol for the post-DN mode and raw wastewater for the pre-DN mode. The minimum anoxic SRT required for both systems to achieve complete DN is shown in Table 6.4. Results from a single sludge pre-denitrification-nitrification process configuration treating wastewater from an organic chemical industry indicate complete denitrification at 22° to 24°C with an anoxic SRT of 5 days.[50] Using a similar process configuration, complete denitrification of coke plant wastewaters has been achieved at 20° to 24°C with an anoxic SRT of 7 days.[51]

The alternate and equally acceptable design methodology is the unit denitrification rate approach. Historically, this is the most widely used approach (for systems operated under non-carbon limiting conditions) and the literature abounds with data. The unit denitrification rate, q_{DN}, is defined as:

$$q_{DN} = \frac{\mu_{DN}}{Y_{DN}} \qquad (9)$$

TABLE 6.4. Minimum anoxic SRT to achieve complete denitrification.

Minimum anoxic SRT required (d)	Temperature (°C)
0.5–1.5	24–26
1–2	14–16
4–5	7– 8

where

q_{DN} = unit denitrification rate, mg NO_3-N removed per mg VSS/d, and
Y_{DN} = denitrifier gross yield, mg VSS grown per mg NO_3-N removed.

The denitrification rate q_{DN} also can be defined as:

$$q_{DN} = \frac{1}{X_{DN}} \frac{dN}{dt} \qquad (10)$$

where X_{DN} is equal to the denitrifier mixed liquor volatile suspended solids concentration, mg/L.

In separate sludge systems, X_{DN} is equivalent to the mixed liquor VSS (X). Thus, design on total biomass concentration is appropriate. Thus for separate sludge systems the denitrification rate can be expressed as:

$$r_{DN} = \frac{1}{X} \frac{dN}{dt} = \hat{r}_{DN} \frac{N}{K_N + N} \qquad (11)$$

where

$\hat{r}_{DN}$ = maximum denitrification rate (d^{-1}), and
K_N = half saturation constant, equal to the N concentration at which r_{DN} is equal to one-half $\hat{r}_{DN}$, mg/L.

At steady state the rate of denitrification in separate sludge systems has been shown to be zero order with respect to nitrate concentration and at a maximum provided there is an excess of available carbon.[54,55] Under these conditions:

$$\frac{1}{X} \frac{dN}{dt} = \hat{r}_{DN} \qquad (12)$$

It also has been shown that the rate of denitrification is not a strong function of SRT in the range of 3 to 9 days and that for separate sludge systems, the following Arrhenius relationship was developed to calculate the denitrification rate:[40]

$$\hat{r}_{DN} = 5.42 \times 10^{10} \, e^{-15300/RT} \quad (13)$$

where

R = Universal Gas Constant (1.987 Cal/ gmole·°C), and
T = temperature (°K).

This relationship is shown schematically in Figure 6.7.

In single or combined sludge systems, X_{DN} is not equal to X, and for design purposes some estimate of the denitrifier population must be made. The most complex system would be the two stage single sludge system designed for both carbonaceous and nitrogenous control.

Under these conditions, X_{DN} can be calculated by:

$$X_{DN} = \frac{Y_{DN}(Ni - Ne)X}{Y_{DN}(Ni - Ne) + Y_N(Ai - Ae) + Y_C(Ci - Ce)} \quad (14)$$

where

$(Ni - Ne)$ = NO$_3$-N removed, mg/L,
$(Ai - Ae)$ = NH$_3$-N removed, mg/L,
$(Ci - Ce)$ = TOC removed, mg/L, in aerobic respiration,
Y_N = nitrifier gross yield (mg VSS/mg NH$_3$-N removed), and
Y_C = heterotrophic gross yield for aerobic respiration (mg VSS/mg TOC removed).

Equation 14 assumes the heterotrophs produced during aerobic respiration are

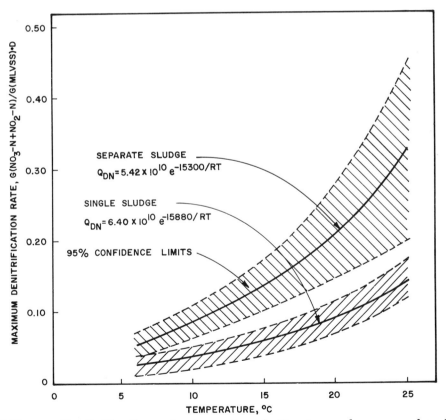

FIGURE 6.7. Denitrification rates in an Anoxic Reactor under non-carbon limiting conditions.

incapable of denitrification. However, it is more likely that a fraction of the heterotrophs, initially grown under aerobic conditions, are capable of nitrate respiration. Consequently, Equation 14 will underestimate the denitrifier population.

However, the design for single sludge systems can be based on total volatile solids, provided the limitations of this approach are realized. It must be emphasized that the denitrification rate of single sludge systems will be lower than for separate sludge systems and will be subject to fluctuations dependent on variations in feed composition.

Pilot-scale evaluation of single sludge nitrogen removal systems indicated no statistical difference of DN rates in the pre- and post-DN modes.[49] Results from this study are shown in Figure 6.7. It can be seen that the results compare favorably with those in the literature and that the rates are roughly 50% of those achieved in separate sludge systems.

Alternate carbon sources. The carbonaceous matter used as the energy source and electron donor for the denitrification reaction can be supplied by organics present in the wastewater, supplemental organic carbon addition, or endogenous respiration.

The most common carbon source quoted in the literature is methanol. However, the price of petrochemical products has all but eliminated the use of methanol as a carbon source for denitrification. Consequently, the use of raw wastewater or other organics present in industrial wastewaters is fast becoming the only economically viable source of organic carbon. Designs based on endogenous respiration are viable. However, the denitrification rates achieved are substantially lower than those achieved under non-carbon limiting conditions with the result that reactor volumes are increased significantly. Generally, the higher the system SRT, the lower will

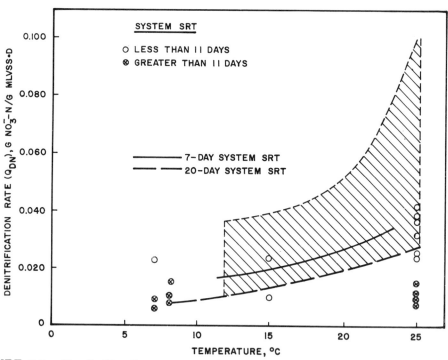

FIGURE 6.8. Denitrification rates using endogenous carbon sources. Crosshatched area shows range of rates as described by EPA.

be the endogenous respiration denitrification rate, as is shown in Figure 6.8.

Extensive research has revealed that wastewater and waste carbon streams are excellent carbon sources for biological denitrification, exhibiting denitrification rates equal to or better than those observed using methanol. Climenhage[54] demonstrated that C_1 to C_5 volatile organic acids, as generated in the manufacture of nylon intermediates, were effective as the carbon source for denitrification. Rates of 0.36 d^{-1} and 0.1 d^{-1} at 20° and 10°C were reported. These are comparable to other results reported for separate sludge systems.

In another study in which brewery wastes were used, DN rates of 0.22 to 0.25 d^{-1} were achieved at 19° to 24°C, compared to 0.18 d^{-1} using methanol.[55] Sutton et al.[49] demonstrated that the organics present in raw wastewater are comparable to methanol as a carbon source for denitrification (Figure 6.9). An evaluation of industrial wastes as carbon sources for denitrification revealed that 27 of the 30 wastes tested exhibited DN rates equal to or greater than those observed using methanol.[56] A listing of the wastes tested with DN rates achieved is shown in Table 6.5. This study also revealed that many of the wastes exhibited substrate consumption ratios equal to or less than those for methanol.

Organic carbon requirements for maximum denitrification rates. Once the quantity of organic carbon required for biological growth and reduction of nitrite, nitrate, and oxygen becomes limiting, denitrification will not proceed at

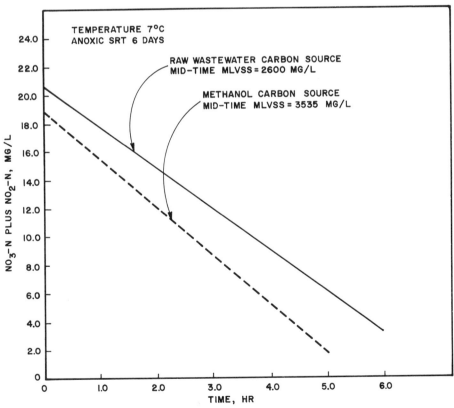

FIGURE 6.9. Denitrification rates with methanol and raw wastewater as carbon sources.

TABLE 6.5.   Denitrification rates using industrial wastes as the carbon source.

Waste	Initial FOC:H	Temp. (°C)	DN rate (d⁻¹)	DN rate relative to MeOH control run on same day	Substrate consumption ratios		Comments[a]
					kg FCOD consumed kg NOT-N removed	kg FOC consumed kg NOT-N removed	
GROUP 1: The following wastes exhibited rates above the 95% confidence interval for methanol:							
Rieder distillery fusel oils	3.19	20.5	0.331	2.38	2.22	0.77	
Pea blanchwater (food processor 'A')	3.27	18.5	0.261	2.08	5.71		
Jordan wines sludge centrate	2.70	20.5	0.207	1.62	7.30	2.28	1
Labatt's brewery spent grain extract	3.18	20	0.197	2.40	5.48	2.46	1
Molson's brewery last runnings	2.53	20.5	0.191	1.49	6.67	1.83	
Molson's brewery wort	4.29	21	0.187	2.27	6.17	1.35	1
McGuinness distillers thin stillage	2.71	21	0.184	1.44	6.07	2.18	1
Mathanol still bottoms (org. manuf. 'A')	1.49	20	0.170	0.86	3.66	0.71	1
National starch process effluent	2.97	18	0.160	1.54	3.26		
Tomato sludge (food processor 'A')	1.72	18	0.160	1.31	2.54	0.80	1
McGuinness distillers fusel oils	3.17	20	0.159	1.29	5.32	1.46	1
Molson's brewery beer	4.16	20.5	0.159	1.41	8.57	2.54	1
Du Pont organic acids waste	2.61	21	0.142	1.29	5.14	1.65	1
Spent sulphite liquor (Can. Int. Paper)	1.77	19	0.137	1.24	3.94	0.79	1
Domtar packaging whitewater	3.72	21	0.137	2.13	5.74	1.48	2,3
Vulcan-Cincinnati methyl fuel	4.07	21	0.135	1.22	6.18	1.83	1,3
Celanese light ends (tray 25)	3.48	21	0.129	1.17	5.23	1.36	
Methanol heads (Ontario Paper Co.)	1.53	18	0.128	1.06	2.45	0.82	
Rieder distillery grape slops	3.21	20	0.125	1.94	5.00	1.42	
Acetic acid waste (Dow Chemical Co.)	1.76	20	0.123	0.62	3.87	1.71	1
Du Pont high boiling organic waste	2.53	19	0.119	1.07	6.02	1.36	1
McGuinness distillers light distillate	9.91	20	0.117	1.61	10.16	1.52	2

GROUP 2: The following wastes exhibited rates within the 95% confidence interval for methanol:

Jordan wines pomace extract	3.43	19	0.112	1.74	5.69	2.6	[1]
Millhaven fibres glycol waste	2.94	20	0.103	1.60	5.98	0.92	
METHANOL CONTROL	2.87	20	0.097	—	5.41	1.17	DN is Mean of 14 Runs
Molson's brewery trub	4.73	20	0.093	1.28	6.40	3.7	
Isopropanol waste (Norwich)	4.40	20	0.090	1.40	3.64	1.82	[2]
Gos and Gris cheese whey	2.50	20	0.084	1.31	9.65	0.91	[1]

GROUP 3: The following wastes exhibited rates within the 95% confidence interval for methanol:

Domter packaging black liquor	2.24	18	0.080	1.25	6.02	1.76	[1,3]
Waste dextrose (Baxter Travenol Labs)	2.65	20	0.071	0.57	8.19	2.57	[1]
Formaldehyde waste (University of Guelph)	3.62	20	0.042	0.37	6.21	1.38	

[a] *LEGEND:* 1) Wastes cause nitrite production. 2) Waste adds to TKN concentration. 3) Waste adds color to clarified effluent.

81

the maximum rate. If the chemical structure of the organic carbon source is known, balanced equations can be developed that will describe the stoichiometric FOC (filtered organic carbon) requirements for nitrogen removal and deoxygenation. McCarty et al.[57] conceived the consumptive ratio, Cr, represented by the relationship:

$$Cr = \frac{\text{Total FOC utilized}}{\substack{\text{FOC required for denitrification} \\ \text{and deoxygenation}}}$$

(15)

A supplemental quantity of organic carbon is required for cell growth and consequently Cr is observed to be greater than unity. As Cr increases, cell synthesis increases. Another method of expressing the organic requirements is simply via the quantity of FCOD (filtered COD) or FOC consumed per unit of nitrogen reduced. This is called the substrate consumption ratio. Monteith et al.[56] determined a mean FOC substrate consumption ratio for methanol of 1.17 (Table 6.5), compared to the stoichiometric requirement of 0.71 for nitrate reduction alone. In this same study the substrate consumptive ratios for 30

industrial wastes were measured. These values ranged from 0.71 to 3.70 g FOC/g N removed.

Pilot-scale studies using the single sludge pre- and post-denitrification-nitrification process configuration for the treatment of domestic wastewater, generated substrate consumption data.[49] The results from the pre-denitrification mode indicated that unless approximately 8 g FCOD were available per g of equivalent NO_3-N reduced, the degree of denitrification would be limited (Figure 6.10). Results from the post-denitrification system, using methanol as the carbon source, revealed that the stoichiometric requirement of about 3 to 4 g FCOD/g N, is adequate to ensure complete denitrification.

In the pre-denitrification mode the organic carbon requirements to ensure no NO_3-N or NO_2-N in the anoxic effluent were determined as a function of feed C:N ratio. These data are shown in Figure 6.11. Unless the feed COD:N ratio was 13 to 15, oxidized nitrogen was evident in the anoxic reactor. These results are similar to those reported by Bridle et al.,[51] who determined the minimum feed FOC/N required to assure complete denitrification of coke plant wastewater

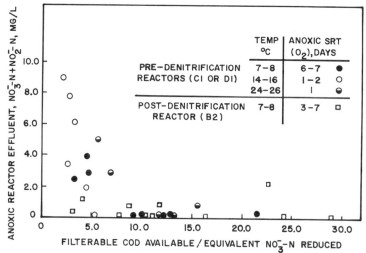

FIGURE 6.10. Results of pilot study showing organic carbon requirements in post- and pre-denitrification reactors.

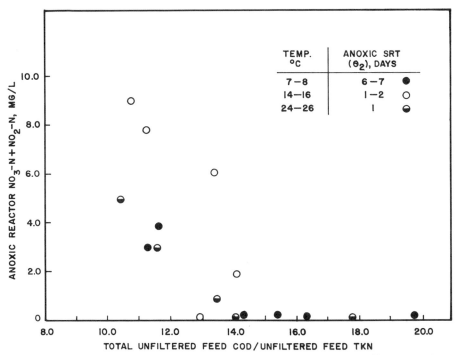

FIGURE 6.11. Results of pilot study showing influent C:N effect on nitrate reduction in pre-denitrification reactor.

(pre-denitrification-nitrification mode) to be 3.5 (Figure 6.12). Stern and Marais[58] and Marsden and Maris[59] operating bench-scale pre-denitrification-nitrification systems, determined the required feed COD:N ratio to be greater than 15. They concluded that the reason for the excessive COD requirement in the pre-denitrification mode is a rapid initial COD removal via adsorption. Full-scale experience at Oswego, N.Y.,[60] and Largo, Fla.,[61] have indicated high levels of nitrogen control with influent BOD_5:TKN ratios of 5 to 6:1. Based on the Oswego COD:BOD_5 ratio of 2.2, this translates to a COD:TKN ratio of 11 to 13:1, supporting the bench- and pilot-scale data. Monteith et al.[56] verified that the feed C/N ratio can affect carbon requirements (Figure 6.13).

Integrated systems for biological nitrogen conversion and control. Integrated biological systems designed for com-

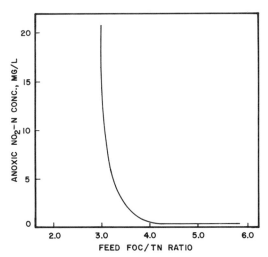

FIGURE 6.12. Effect of feed FOC/TN ratio on denitrification.

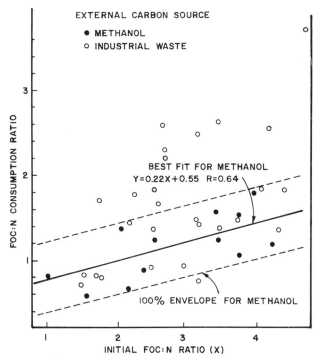

FIGURE 6.13. **Substrate consumption ratio versus initial FOC:N ratio.**

bined nitrogen conversion and removal can be either separate or single sludge systems, operated in either the pre- or post-DN mode. Separate sludge systems operate at higher unit removal rates and consequently require lower reactor volumes. However, this advantage is normally overshadowed by the advantages attributed to single sludge systems, namely: reduction of clarification requirements, improved sludge properties, reduced need for pH control, reduced aeration requirements, and reduced carbon requirements (pre-DN mode).

Single sludge systems for combined carbon oxidation, nitrification, and denitrification seem to be the most cost-effective system for nitrogen control,[51,62-64] provided effluent requirements are compatible with process capability. Recent studies have indicated that the anoxic or aerobic periods created in these process configurations do not affect nitrification kinetics, and no difference in denitrification performance is

evident whether the systems are operated in the pre-DN or post-DN mode.[49] The advantage of the pre-denitrification mode is that the wastewater organics can serve as the electron donor for the denitrification reaction. A study conducted for the City of Penticton, British Columbia, Canada, indicated the most cost-effective process configuration to achieve nitrogen control was the single sludge pre-denitrification-nitrification system. Design parameters generated in this study were used to develop cost comparison data (Table 6.6).[64]

Design example. The following 10 000 m³/d (2.65 mgd) design example is provided to illustrate how the design methodologies are used. Three process configurations will be addressed: separate sludge post denitrification; single sludge post denitrification; and single sludge pre-denitrification. The three process configurations are depicted schematically in Figure 6.14. Raw wastewater

TABLE 6.6. Economic comparison of biological nitrogen removal processes.

	Activated sludge	Nitrification	Two-sludge post-DN	One-sludge post-DN	One-sludge pre-DN
Present worth[a]					
Capital	$2 618 000	$3 268 000	$4 802 000	$4 180 000	$4 217 000
Present worth of annual O&M	1 103 000	1 679 000	3 366 000	2 996 000	2 567 000
Total present worth	$3 721 000	$4 947 000	$8 168 000	$7 176 000	$6 784 000
Comparative rating	0.55	0.73	1.20	1.06	1.00
Annual cost					
Amortized capital[a]	$ 318 000	$ 397 000	$ 583 000	$ 508 000	$ 512 000
O&M	134 000	204 000	409 000	364 000	312 000
Total annual cost	$ 452 000	$ 601 000	$ 992 000	$ 872 000	$ 824 000
Comparative rating	0.55	0.73	1.20	1.06	1.00

[a] Based on a 10.5% discount/interest rate for 20 years.
NOTE: All costs are in January 1979 dollars.

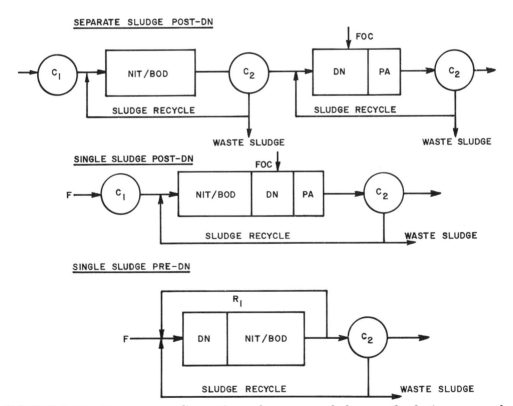

FIGURE 6.14. Process configurations for suspended growth design example. Nomenclature: C_1 = primary clarifier; C_2 = secondary clarifier; NII/BOD = nitrification/BOD removal; DN = Denitrification; PA = post aeration; R_1 = mixed liquor recycle; and, FOC = supplementary carbon addition.

characteristics, primary effluent, and required effluent quality are shown in Table 6.7. The expected reactor temperature is 10° to 25°C, with effluent quality to be met at 10°C.

Separate sludge post-DN: It is assumed that the first stage aerobic reactor after primary clarification achieves complete nitrification as defined by a TKN less than 3 mg/L and TSS of 20 mg/L in the effluent. The soluble BOD_5 will be equal to or less than 10 mg/L in the effluent. The DO is 3 mg/L. Cellular synthesis will remove a portion of the influent TKN which can be determined from the sludge yield in the first stage and the organic nitrogen content of the VSS.

Based on a yield of 0.55 kg TSS/kg BOD_5 (lb TSS/lb BOD_5) removed, an 80% volatile fraction in the TSS, and a 9% organic nitrogen content in the VSS, the cellular synthesis TKN removal is calculated as follows:

Synthesis TKN
$= [(160$ mg/L TBOD-10 mg/L SBOD$) \times$

0.55 mg TSS/mg $BOD_5 - 20$ mg/L TSS$]$

$$\times \left(\frac{80\% \text{ volatile}}{100\%}\right)\left(\frac{9\% \text{ Org-N}}{100\%}\right)$$

$$= \left[(160 - 10)0.55 - 20)\right]\left(\frac{80}{100}\right)\left(\frac{9}{100}\right)$$

$= 4.5$ mg/L TKN in waste sludge, and
$= 1.4$ mg/L TKN in effluent suspended
solids

Effluent solids will contain 20 mg/L TSS at 80% volatile and 9% organic nitrogen or 1.4 mg/L TKN. Soluble TKN then will be equal to or less than 3.0 − 1.4, or 1.6 mg/L. Effluent standards based on TKN are much stricter than NH_4^+-N effluent criteria. At 20 mg/L TSS and a 3 mg/L TKN effluent criteria, it will be necessary to reduce the effluent NH_4^+-N to about 1.0 mg/L. This value should be used to calculate the nitrification volume.

Based on the equivalent of 4.5 mg/L

TKN in the waste sludge and 3 mg/L in the final effluent, $26 - 4.5 - 3.0$ mg/L or 18.5 mg/L must be oxidized to NO_3-N. The equivalent NO_3-N value of 3 mg/L DO influent to the DN stage is (0.33) (3 mg/L DO), or 1 mg/L NO_3^--N equivalent. Thus, the total equivalent NO_3^--N feed to the DN reactor, for calculation of the supplementary carbon addition (FOC), is 19.5 mg/L, less the effluent NO_3^--N of 5 mg/L.

Assuming an alkalinity removal of 7.1 mg $CaCO_3$ alkalinity per mg of NO_3-N produced, the first stage residual alkalinity would be $[200 − (7.1)(18.5)]$ or 69 mg/L. This may be marginal with load variations for assuring effective first-stage nitrification and could inhibit DN. Alkalinity supplement would be made to the first stage to utilize the higher pH in both stages. It is recommended that a minimum pH of 7.0 be maintained. DN will increase the pH in the second stage because of 2.9 mg $CaCO_3$ alkalinity produced per mg NO_3-N reduced to N_2. Total increase in alkalinity in the DN reactor would be 39 mg/L $[(2.9)(18.5 − 5.0)]$.

The maximum denitrification rate achievable for the separate sludge system is calculated from Equation 13.

$$\hat{q}_{DN} = 5.42 \times 10^{10}\ e^{-15300/RT}$$

For 10°C, $\hat{q}_{DN} = 0.083$ kg/kg·d NO_3-N/ VSS

The NO_3-N load to the DN reactor is 185 kg/d. Therefore, the biomass required is:

$$\frac{185}{0.083} = 2229 \text{ kg VSS}$$

(assume at 80% volatile)

Designing the reactor for mixed liquor VSS of 2500 mg/L (3125 mg/L TSS) yields a reactor volume of:

Reactor volume =

$$\left(\frac{2229}{2500}\right)(1000) = 891 \text{ m}^3$$

with a hydraulic retention time of:

$$HRT = 2.2 \text{ hours.}$$

TABLE 6.7. Feed characteristics/effluent criteria.[a]

Parameter	Raw wastewater mg/L	kg/d	Primary effluent mg/L	kg/d	Final effluent mg/L	kg/d
Total BOD$_5$	240	2 400	160	1 600	20	200
Filtered BOD$_5$	80	800	80	800	—	—
TKN	30	300	26	260	3	30
TSS	250	2 500	120	1 200	20	200
Alkalinity	200	2 000	200	2 000	—	—
NO$_3$-N	0	0	0	0	5	50

[a] Average flow is 10 000 m^3/d.

No additional safety factor was used, as the design is based on the minimum temperature expected, and the effluent quality requirement is only 5 mg NO$_3$-N/L. The cold weather safety factor could be considered to be 37% [(18.5/18.5 − 5)(100)]. Warm weather conditions permit a higher degree of denitrification. However, the DN rate and reactor size were based on the use of methanol as a carbon source. If another organic compound is to be used, it may be necessary to reduce the DN rate and enlarge the basin.

If a denitrifier net TSS yield coefficient of 0.8 is assumed, the quantity of biomass produced per day is equal to:

Biomass/d = 116 kg TSS/d (0.8 × 145) + 200 kg TSS/d

The 200 kg/d TSS comes from the nitrification clarifier carryover. This calculation assumes no additional endogenous respiration in the DN basin.

Therefore, at equilibrium, the SRT would be:

$$SRT_{DN} = \frac{(2229)}{(116 + 200)0.8} \quad \frac{1}{}$$

$$= 8.8 \text{ days}$$

Thus, from an SRT approach, the design is adequate. The carbon requirements for the post DN system are approximately 3 kg methanol/kg equivalent NO$_3$-N removed. For this system, approximately 435 kg/d methanol (3 × 145 kg equivalent NO$_3$-N) would be required. The cost (1980) would be approximately $49 600 per annum for methanol ($0.31/kg).

The net yield for the first stage biological system is assumed to be 0.55 kg TSS/kg BOD$_5$ (lb TSS/lb BOD$_5$) removal of which 10 mg/L SBOD$_5$ and 20 mg/L TSS passed into the second stage. Thus, the net winter sludge production (10°C) based on a 20/20 final effluent is as follows:

Stage	TSS Produced kg/d	TSS Effluent kg/d	Waste Sludge kg/d
Primary Clarification	1 300	1 200	1 300
1st (N)	880	200	680
2nd (DN)	116		116
Carryover	200	200	
Solids Production			2 096

In the summer months, the waste sludge would be 10 to 15% lower if the same MLSS was maintained (due to higher SRT).

Single sludge post-DN. The organic load to the DN reactor of the single sludge system will be identical to the separate sludge system; that is, an NO_3-N feed of 18.5 mg/L or a mass loading of 185 kg/d. For the single sludge system;

$$q_{DN} = 6.40 \times 10^{10}\ e^{-15580/RT}$$

and for 10°C,

$$q_{DN} = 0.035\ kg/kg \cdot d\ NO_3\text{-}N/VSS.$$

Thus, the biomass required is:

$$\frac{185}{0.035} = 5286\ kg\ MLVSS.$$

Design of the DN MLVSS at 2500 mg/L yields a reactor volume and hydraulic retention time of $2\,114$ m^3 and 5.1 hours, respectively. For the same reasons outlined for the separate-sludge system, no safety factor need be applied for the single-sludge post DN system, and methanol requirements will be the same as for the separate sludge system. Sludge production in this flowsheet will be similar to the separate sludge process. Thus, the DN equilibrium SRT can be calculated to be:

$$SRT\ (DN\ Equil.) = \frac{5286}{0.8(880 + 116)}$$

$$= 6.6\ d$$

Single sludge pre-DN. In the single sludge pre-DN system, primary clarification is not required. It frequently is purposely deleted to increase the available quantity of biodegradable organics for reducing the NO_3-N. However, with the omission of the primary clarifier, it is necessary to adjust for the presence of non-biodegradable VSS present in the MLSS, as well as the increased TKN entering the reactor.

Based on a net yield of 0.8 kg TSS at 75% volatile per 1.0 kg BOD$_5$ removed and an organic nitrogen content of 7% o MLVSS, the residual soluble TKN is:

Soluble TKN =
$$30\ mg/L - [(240 - 10)(0.8)](0.75)(0.07$$

$$= 20.3\ mg/L$$

Because the effluent soluble TKN is 1.(mg/L and 5 mg/L NO_3-N, the quantitγ remaining to be nitrified is 13.9 mg/I TKN or 139 kg/d.

The DN rate for systems withou^{t} primary clarifiers is 80% of the rate where primary clarification is used Thus, the q_{DN} is [(0.035)(0.80)], or 0.024 kg/kg$\cdot$d NO_3-N/MLVSS.

The quantity of MLVSS in the DN portion of the reactor is equal to:

$$\frac{139\ kg/d}{0.028\ d^{-1}} = 4964\ kg\ MLVSS$$

With a MLVSS of 2500 mg/L, the DN volume is 1986 m^3, equivalent to a raw wastewater detention time of 4.8 hours. Like the single sludge post DN, the SRT$_{DN}$ of this process is difficult to define and is not directly comparable to a two-sludge design.

The minimum carbon requirements for the pre-DN flowsheet have been estimated at 14 kg COD/kg NO_3-N equivalent. Assuming a COD/BOD$_5$ ratio of 2.0, the feed COD would be 480 mg/L. The nitrate equivalent of the DO in the recycled mixed liquor is:

$$3 \times 0.33 \times 5\ recycles =$$
$$5\ mg\ NO_3\text{-}N/L\ equivalent$$

Therefore, the total nitrate equivalent for DN is:

$$5 + 13.9 = 18.9\ mg/L\ equivalent\ NO_3\text{-}N$$

The COD consumed will be equal to 18.9 mg/L, or 189 kg/d NO_3-N equivalency. A total of 79 mg/L COD will be reduced in the DN portion of the biological reactor. The balance of the COD, $4\,001$ kg/d ($4\,800 - 799$), is discharged from the DN basin into the nitrification basin.

Although the net yields of carbonaceous BOD_5, nitrifiers, and denitrifiers could be determined individually, the level of refinement is not warranted when working with a single sludge pre-DN system without primary clarifiers. If pilot studies have not been conducted, it is suggested that the net yields be determined from comparable data using the relationships:

$$Y = a\,BOD_{5R} - b\,MLSS$$

and

$$\frac{1}{SRT} = a(F/M)_R - b$$

where:

a = cell yield or synthesis factor, lb/lb,

b = endogenous decay coefficient, day^{-1},

$BOD_{5R} = BOD_5$ removed, mg/L, and

$(F/M)_R = BOD_{5R}/MLSS \cdot day$

If the coefficients a and b are unknown and the waste characteristics are consistant with the criteria set forth, Figures 14-15 and 14-16 from the Joint ASCE-WPCF Wastewater Treatment Plant Design manual (1982) can be used to determine the net yield. The input to the curve would be SRT_n, SRT_{DN}, and

wastewater temperature. The "winter-summer" average net yield at a SRT of 25 days will be about 0.8 lb TSS/lb BOD_5 removed. Thus, the waste sludge quantity would be:

$$Y_w = (2\,400 - 100)0.8 - 200$$
$$= 1\,640\ kg/day$$

The SRT of the DN compartment would be 4 964/1 640 + 200 or 2.7 days. This is a low SRT because of the use of the total net yield of carbonaceous BOD_5, nitrification, and denitrification. Design should be adequate at the SRT_{DN} and HRT of 4.8 hours. The significant design data generated from these examples are summarized in Table 6.8.

The design engineer should ensure that there is careful consideration of the values of q_{DN}. Denitrification technology will continue to evolve. Engineering judgment should consider cost/benefits, both in the settling of effluent standards as well as the process design. Different standards for cold weather and warm weather are becoming common and reflect good judgment in establishing standards for BOD_5, N, and DN.

AIR STRIPPING OF AMMONIA

Ammonia nitrogen can be removed from water by the volatilization of

TABLE 6.8. Design summary for the three process configurations.

Parameter	DN process configuration		
	Separate sludge post-DN	Single sludge post-DN	Single sludge pre-DN
q_{DN} 10°C (d^{-1})	0.083	0.035	0.028
DN reactor volume (m^3)	891	2 114	1 986
DN HRT (h)	2.2	5.1	4.8
Calculated DN SRT (d)	7.2	5.1	2.7
Alkalinity required	Yes	Possibly	No
MeOH cost ($/yr)	49 600	49 600	0
Post aeration	Yes	Yes	No
No. clarifiers	3	2	1
Sludge generated			
Primary—kg/d	1 300	1 300	0
Secondary—kg/d	828	828	1 640
Total—kg/d	2 128	2 128	1 640

gaseous NH_3^0 into the air. The process is simple in concept and reliable, but it has serious drawbacks that make it relatively expensive: the very low vapor pressures of NH_3^0 at ambient temperature, and neutral pH.

The rate of transfer is enhanced by converting most of the ammonia to gaseous NH_3^0 at a high pH (usually 10.5 to 11.5), but this in turn favors the absorption of CO_2 from the air and possible carbonate scaling. These difficulties limit the practical application of air stripping to special cases, such as the need for a high pH for another reason, the availability of waste heat from another process, and ammonia concentrations so gross as to be toxic to biological methods of treatment and enhance the economics of air stripping. Air stripping also may be used to remove many hydrophobic organic molecules economically.

Process fundamentals. In a wastewater, ammonium ions exist in equilibrium with ammonia as shown by Equation 16.

$$NH_3^0 + H_2O \rightleftharpoons NH_4^+ + OH^- \quad (16)$$

At a pH level of 7, only ammonium ions (NH_4^+) in true solution are present. At a pH of 12, ammonia (NH_3^0) exists in solution as a dissolved gas. In the range of pH 7 to 12, ammonium ions and ammonia gas co-exist together in proportional percentages as a function of the pH. As the pH of the wastewater is increased above 7, the equilibrium shown in Equation 16 is shifted to the left in favor of ammonia gas, which can be removed from the liquid by aeration. The relative percentages of ammonium ions and ammonia at different pH levels are shown in Figure 6.15 for different temperatures.[65]

To determine the amount of base needed to adjust the pH of a wastewater, it is possible to use data presented in the form of Figure 6.16.[66] This figure presents typical data indicating the quantity of lime required to adjust

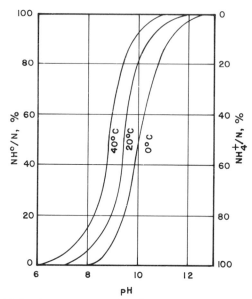

FIGURE 6.15.   Distribution of ammonia and ammonium ion with pH and temperature.

the pH to 11 as a function of the wastewater alkalinity.

After the ammonium ion has been converted to ammonia in solution, there are two major factors other than temperature that affect the rate of transfer from the wastewater to the surrounding atmosphere. These two factors are the rate of renewal of the air-water interface and the driving force resulting from the difference in ammonia concentration in the water and the surrounding air. The rate of renewal of an air-water interface is minimal while the surface film is being formed and the release of ammonia gas would be greatest during this instance. It has been found that during the process of forming water droplets, the surface tension is at a minimal value. After the water droplet has been formed, gas transfer becomes negligible. Therefore, repeated droplet formation, rupture, and reformation greatly assist ammonia stripping operations with regard to minimizing surface tension. To

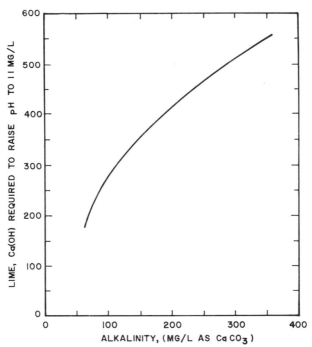

FIGURE 6.16. Lime required to raise the pH to 11 as a function of the raw wastewater alkalinity.

maintain a sufficient driving force between the liquid and air phases, it is necessary to minimize the concentration of ammonia in the air phase. This may be accomplished by circulating large quantities of air rapidly through the water droplets. These same principles of droplet formation and reformation and the necessity of large gas/liquid requirements are applied to conventional cooling towers and explain the adaptability of these towers to the ammonia stripping process.

The next step in evaluating the ammonia stripping process is to determine the quantity of air required to remove the ammonia from solution. Because it is assumed that a stripping tower will accomplish the ammonia removal, the materials balance sketch shown in Figure 6.17 can be described by Equation 17 as:[66]

$$G(Y_1 - Y_2) = L(X_1 - X_2) \quad (17)$$

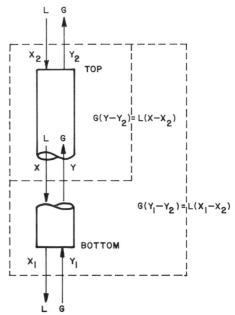

FIGURE 6.17. Materials balance for a counterflow stripping tower.

where

G = molar gas flow, kg·mol/s
L = molar liquid flow, kg·mol/s
$Y_{1,2}$ = mole fraction in gas, in and out, and
$X_{1,2}$ = mole fraction in liquid, in and out.

This equation is based on a mass balance of input and output and is valid regardless of the internal conditions that control the process. Fortunately, the internal equilibria are governed by Henry's Law for which adequate knowledge exists and that will not be described.

The curves shown in Figure 6.18[66] were calculated using Henry's Law at one atmosphere pressure. By using Figure 6.17 and Equation 17, the theoretical air requirements per unit of water can be calculated. By assuming that the water leaving the air entering the bottom of the stripping tower is free of ammonia, Equation 18 can be expressed as

$$\frac{G}{L} = \frac{X_2}{Y_1} \qquad (18)$$

where G/L represents the ratio of air to wastewater required to establish the ammonia equilibrium at a specified temperature and pressure. It must be assumed that the ammonia in the air leaving the stripping tower is in equilibrium with the wastewater entering the tower. Therefore, under these conditions the slope of the line in Figure 6.18 would indicate the minimum air to liquid ratio at a given temperature. To illustrate this approach, assume a temperature of 20°C. The gas/liquid ratio would equal the ratio of X/Y or:

$$\frac{X}{Y} = \frac{\dfrac{\text{moles NH}_3}{\text{moles H}_2\text{O}}}{\dfrac{\text{moles NH}_3}{\text{moles air}}} = \frac{0.02 \text{ moles air}}{0.015 \text{ moles H}_2\text{O}}$$

$$= 1.33 \text{ moles air/mole H}_2\text{O}$$

By conversion, 1.33 moles of air will equal 29.88 L (1.055 cu ft) and one mole

of water equals 0.018 L (0.0047 gal) as shown in the following computations:

$$1.33 \text{ moles} \times 29 \frac{\text{g}}{\text{mole}} \times \frac{T}{454} \frac{\text{g}}{\text{lb}}$$

$$\times \frac{T}{0.0808} \frac{\text{lb}}{\text{cu ft}} = 1.055 \text{ cu ft } (29.88 \text{ L})$$

$$1.0 \text{ mole} \times 18 \frac{\text{g}}{\text{mole}} \times \frac{T}{454} \frac{\text{g}}{\text{lb}}$$

$$\times \frac{T}{8.34} \frac{\text{lb}}{\text{gal}} = 0.0047 \text{ gal } (0.018 \text{ L})$$

The required air liquid ratio is, therefore:

$$\frac{G}{L} = \frac{1.055 \text{ cu ft}}{0.004\ 76 \text{ gal}}$$

$$= 223 \text{ cu ft/gal } (1660 \text{ m}^3/\text{m}^3).$$

This same approach can be made for different temperatures with the results shown in Figure 6.19.[66] These theoretical results assume the process to be 100% efficient, which is obviously not true in actual practice. A suggested rule of thumb is to multiply the theoretical calculation by a factor of 1.5. The application of this factor is also shown in Figure 6.19.

Design methodology. The selection of the optimal design criteria for a stripping tower generally first requires evaluation of the wastewater quality and the need for treatment prior to stripping. The NH$_3$ removal desired then is used to determine unit process efficiency and tower configuration. The following sections deal with the selection of tower configuration, packing and depth, materials, and design characteristics. A discussion of the advantages and disadvantages of the process, as well as the costs associated with the design and operation, also are included.

Schematic arrangement of stripping process. The air stripping process has been used to treat raw waste, anaerobic digester supernatant, and secondary biologically-treated effluent.

Possible flow-sheets for the process, including recovery and reuse of the lime

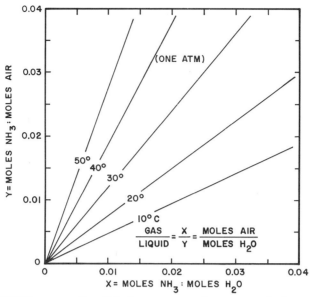

FIGURE 6.18. Equilibrium curves for ammonia in water.

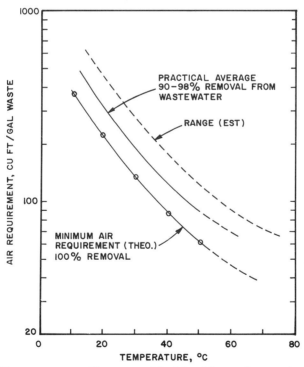

FIGURE 6.19. Temperature effects on Air/Liquid requirements (G/L) for ammonia stripping. NOTE: cu ft/gal $\times$ 7.48 $=$ m^3/m^3.

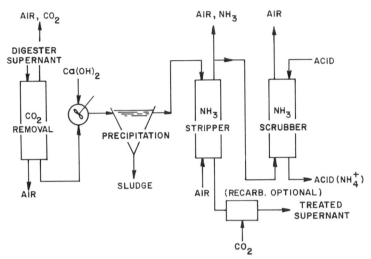

FIGURE 6.20. **Flowsheet for the treatment of digester supernatant for nutrient removal.**

for pH adjustment and phosphorus removal, are shown in Figures 6.20, 6.21, and 6.22. Of course, with industrial raw wastes, the phosphorus removal may not be relevant and caustic sodium hydroxide might be employed for pH adjustment.

Equipment. Because of the high air:liquid requirements needed for ammonia stripping, it has been found that conventional cooling towers provide the

most suitable means of providing the desired air volume. The two most common towers are the "countercurrent" and the "crossflow" (Figure 6.23).

Other possible methods of stripping ammonia from solution include diffused air and mechanical surface aerators.

Helix-type aerators also have been used with varying degrees of success.

The major advantage of these other aerators is their ability to maintain a higher temperature of the wastewater

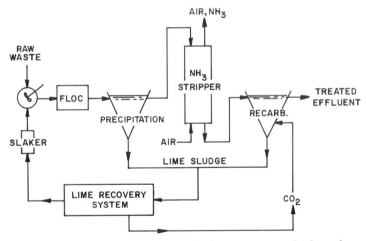

FIGURE 6.21. **Flowsheet for the removal of nitrogen and phosphorus from raw wastewater.**

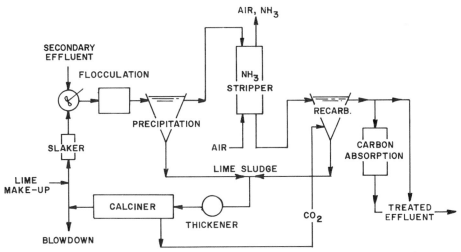

FIGURE 6.22. Flowsheet for the removal of nitrogen and phosphorus from secondary effluent.

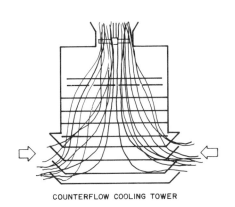

COUNTERFLOW COOLING TOWER

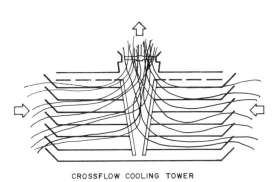

CROSSFLOW COOLING TOWER

FIGURE 6.23. Conventional cooling towers commonly used for ammonia stripping.

during stripping. However, these devices are not capable of supplying the extremely high air quantities as economically as cooling towers.

Process cost data. The operating capital costs of the air stripping process have escalated rapidly during the last decade. The total average cost was about $0.006/m³ treated ($0.017/1000 gal) for the 0.16 m³/s (3.75 mgd) South Tahoe crossflow tower facility.[31,67]

Costs of owning and operating a stripping tower will depend on the design and cost information obtained from the equipment manufacturer. In general, the equipment cost components include: fans; water pumps to deliver the water to the top of the tower; the structure as a function of the volume; the packing as a function of the volume of the tower; and the exterior covering and electrical installation.

The principal operating and maintenance costs are electrical power for pumping water and air, operating and maintenance personnel, and chemicals if required.

Application and results. The very few ammonia-stripping towers designed for treatment of wastewaters generally have been constructed for the treatment of low-strength, municipal wastewaters. Prather[68] has shown that very efficient ammonia removals from solution can be attained through the air-liquid contact provided by the drop formation and re-formation in a stripping tower. By passing the wastewater through a closely packed aeration tower, ammonia-nitrogen removals by air stripping were found to be greater than 90% at any pH above 9.0, while supplying $3\,591$ m^3 of air/m^3 (480 cu ft of air/gal) of wastewater. In other studies with petroleum refinery wastes,[69,70] Prather found the efficiency of ammonia removal to increase from 34 to 85% at pH values of 9.4 and 10.5, respectively, with an air flow rate of $2\,245$ m^3 of air/m^3 (300 cu ft of air/gal) of wastewater. Rohlich[71] observed increasing ammonia removals as the pH was increased from 8.0 to 11.0. No significant increases in efficiency were noted in increasing the pH from 11.0 to 12.0. Ammonia-nitrogen removals also were found to improve as the air-liquid loading increased from 40 to

$3\,345$ m^3/m^3 (5.3 to 447 cu ft/gal). Removals of 92% were observed.

The most publicized application of ammonia stripping with municipal wastewaters is probably the work conducted at Lake Tahoe and reported by Culp and Slechta.[65] A countercurrent, stripping tower 7.62 m (25 ft) high, 1.83 m (6 ft) long and 1.22 m (4 ft) wide was used to examine the effects of surface hydraulic loading rate and air supply on ammonia removal at various depths in the tower. The results of these studies are shown in Figures 6.24, 6.25, and 6.26.

The greater efficiency of ammonia removal with increased depth of tower packing is shown in Figure 6.26 for specific air/liquids loadings. The minimum air/liquid loading required to obtain 90% removal is approximately $2\,245$ m^3 of air/m^3 (300 cu ft of air/gal) of wastewater. This value is in line with the previously mentioned theoretical value of 223 cu ft/gal ($1\,660$ m^3/m^3) multiplied by the stripping factor of 1.5. Figure 6.25 indicates that the deeper depths showed little effect of hydraulic flow rate, recorded as gpm/sq ft, at less than

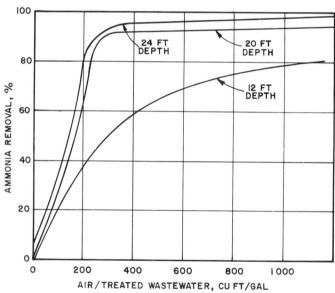

FIGURE 6.24. Effect of tower depth on ammonia removal. **NOTE: cu ft/gal × 7.48 = m^3/m^3; ft × 0.30 = m.**

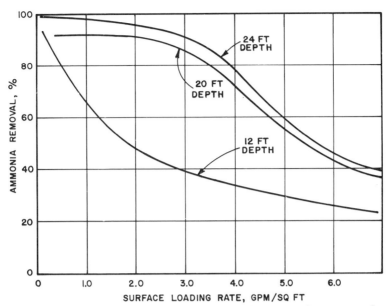

FIGURE 6.25. Effects of hydraulic loading on ammonia removal at various depths. NOTE: gpm/sq ft $\times$ 0.68 = L/m$^2 \cdot$ s.

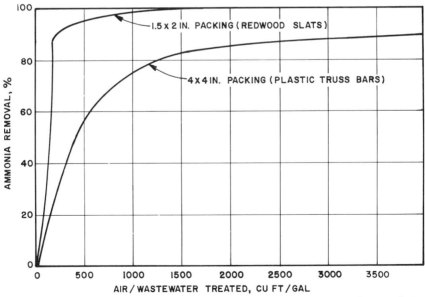

FIGURE 6.26. Effects of packing spacing on ammonia removal. Depth is 7 m (24 ft). NOTE: cu ft/gal $\times$ 7.48 = m^3/m^3; in. $\times$ 25.4 = mm.

3 gpm/sq ft (2 L/m^2·s). At hydraulic loadings greater than 3 gpm/sq ft, sheeting of the water was observed, subsequently reducing tower efficiency. It is recommended that the hydraulic loading to a tower be maintained within the range of 1 to 3 gpm/sq ft (0.67 to 2 L/m^2·s) corresponding to 500 to 1 500 lb/hr/sq ft (2 445 to 7 325 kg/m^2·h). The importance of droplet formation is emphasized in Figure 6.26, which shows the ammonia removal at different air/liquid loadings for two different size packings. The 38 × 51-mm (1.5 × 2 in.) packing has 2.67 times more surface for droplet formation and coalescing than does the 102 × 102-mm (4 × 4 in.) packing. The results in Figure 6.26 also indicate that the efficiency of ammonia removal decreases more rapidly with increased hydraulic loading rates with the 102 × 102-mm packing than with the 38 × 51-mm packing. Visual observation indicated that the smaller-spaced packing resulted in less sheeting of the liquid.

System advantages and disadvantages. Aside from economical considerations, the major advantage of ammonia stripping is the ability to control the process for selected ammonia removals. However, several disadvantages are inherent to the air-stripping process through conventional cooling towers.[72]

Disadvantages of air-stripping in conventional cooling towers.
Cold Weather Operation—Ammonia solubility greatly increases with lower temperatures. Inability to operate tower at wet bulb temperatures less than 0°C (32°F) arises. Fogging and icing problems occur.

Deposition of calcium deposits occur within tower from pH adjustment with lime. This is another problem.

Air Pollution Problems—Ammonia reaction with sulfur dioxide forms aerosol. Wood packing deteriorates because of high pH of wastewaters.

In northern climates with extremely low winter temperatures, several problems may be encountered because of freeze-up of the tower when the wet-bulb temperature is less than 0°C (32°F). Also, a difficulty in the transfer of ammonia from solution is experienced as a result of increased solubility at the lower temperatures. In the larger facilities, a warm influent waste temperature will be highly vaporized causing extreme fogging conditions with the possibility of forming ice layers on nearby roads or buildings.

One of the major disadvantages of the stripping tower is the scaling problem. Several approaches have been proposed, and used, to alleviate the carbonate scaling problem. These include:

• Increasing the performance of the initial lime precipitation process by adding soda ash for non-carbonate hardness removal;

• Using very open, easily cleaned plastic packing. Greater packing volumes or heated air are required to compensate for the loss in mass transfer efficiency;

• Stripping the CO_2 from the air prior to using it in the ammonia stripping tower. This can be accomplished in relatively small packed towers where the air is brought into contact with recycled, high pH water;[70]

• Recycling the exhaust air back into the ammonia-stripping tower after first removing the ammonia in an absorption tower.[31,73] The absorption tower contains recycled water at a pH sufficiently low to convert most of the ammonia to NH_4^+ at the air-water interface. The CO_2 initially present in the air is quickly lost in the stripper column, alleviating the carbonate scaling problem. Other advantages are the ability to recover the ammonia from the absorber water; the elimination of ammonia air pollution problems; and the buildup of the air temperature to that of the wastewater during cold-weather operation.

• Operating at less than optimal pH levels in the ammonia-stripping tower.[73] This inhibits CO_2 transfer by decreasing the free hydroxide concentration, and the lime requirements are reduced. The mass transfer efficiency of the packing is also reduced, but not necessarily to the degree resulting from the approach described above.

Most of these remedial measures will increase the unit cost of the air-stripping process.

The NH_3 transferred to the atmosphere from single-pass stripping towers might pollute the air. This apparently has not been a serious problem for the ammonia concentrations commonly encountered in domestic wastewaters.[31] Many hydrophobic compounds also are stripped along with the ammonia, and some of these could re-pollute natural waters located some distance from a stripping tower.

ION EXCHANGE

Theory. The ion exchange process is a reversible chemical reaction in which an ion in solution interchanges with an ion on the ion exchange media (a solid). This exchange produces no significant change in the structure of the ion exchange media. The exchange phenomenon generally can be modeled by a mass action equilibrium reaction of the following type:

$$b\ A^{+a} + a\ BZ_b, \text{ and } A_a + a\ B^{+b} \quad (19)$$

where A and B are ions with valences a and b, respectively, and Z is the exchange site in the media.

In the field of municipal wastewater treatment and nitrogen removal, the ion to be removed from the waste stream is ammonium, NH_4^+. The ion that the ammonium displaces varies with the nature of the solution used to regenerate the bed. (Regeneration is the process of removing the accumulated NH_4^+ from the ion exchange media so that the media can be reused.) If a sodium solution is used to regenerate the beds, Equation 19 can be rewritten as:

$$NH_4^{+1} + NaZ \quad NH_4Z + Na^{+1} \quad (20)$$

Media selection. It is important to select the proper ion exchange media to optimize the ion exchange process. When selecting the media, consideration must be given to stability and abrasion resistance, selectivity for the NH_4^+ ion to be removed relative to other ions present, stability in relation to the pH of the liquid stream, availability, and cost.

For the removal of ammonium ions from wastewater, experience has indicated that clinoptilolite, a naturally occurring zeolite, is one of the best ion exchange media in terms of the above considerations. In addition to having a greater affinity for ammonium ions than other synthetic and natural media, it is relatively cheap when compared to synthetic media. In addition, clinoptilolite occurs in several extensive deposits throughout the western U.S.

Clinoptilolite properties. Clinoptilolite is preferentially selective for ammonium ion over most ions present in municipal wastewater treatment. The order of preference for ion capture is as follows:

$$K^+ > NH_4^+ > Na^+ > Ca^{++} > Fe^{+3} > Mg^{++}$$

Because potassium concentrations are relatively small when compared to ammonium ions in municipal wastewaters, potassium is not generally a hindrance to removal of ammonium ions, and ammonium therefore is the "preferred" ion exchanged in clinoptilolite.

The amount of NH_4^+ that can be absorbed from a waste stream is dependent on the NH_4^+ concentration delivered to the exchange beds and the allowable effluent NH_4^+ concentration. The total exchange capacity of clinoptilolite in a pure ammonium ion solution is approximately 1.6 to 2.0 meq $NH_4 -N$/g of clinoptilolite. For municipal wastewaters containing typical concentrations of competing cations, the total ammonium ion exchange capacity is approximately 0.4 meq/g, or 7.2 mg of $NH_4 -N$/g of clinoptilolite. These values occur when the clinoptilolite becomes saturated. The

99

value obtained in operating systems where the effluent ammonium concentration is limited to 2 to 6 mg/L is usually closer to 0.3 meq/L.

Clinoptilolite used in most ion exchange specifications is 20 by 50 mesh. With the use of 20 by 50 mesh clinoptilolite, there is little variation in bed performance between flow rates of 7.5 BV/hr and 15 BV/hr. Ion exchange capacity of clinoptilolite has been shown to be optimum in a pH range of 4 to 8. Outside this range, ammonium removal capacity is greatly reduced.[73,74]

In general, the bed depth of the clinoptilolite is related to the amount of ammonium removal that can be accomplished. Pilot studies have shown that a 0.91-m (3-ft) bed depth will remove 0.25 meq of NH_4^+/g of clinoptilolite, while a 1.83-m (6-ft) bed depth will remove approximately 0.32 meq/g.[75]

The backwash rate is chosen to obtain proper cleaning of the media while also avoiding excess fluidization of the bed. This fluidization can cause loss of media in the effluent. Typical values used for backwashing are 4.1 to 5.4 $L/m^2 \cdot s$ (6 to 8 gpm/sq ft).

Process considerations for clinoptilolite media.

Pretreatment. To prevent excessive head loss in the bed, a flow stream low in suspended solids is required. The actual value of suspended solids that can be tolerated in the waste stream will depend on how much head loss can be accepted.

Organic fouling. In general, organic growths are not a problem as backwashing and regeneration with the brine solution usually removes any growths. Care should be taken to design ion exchange beds so that no dead spaces occur that may lead to septic areas in the beds. This consideration is especially true in the case of horizontal beds.

Competing cations. Although clinoptilolite is selective for ammonia relative to most other cations found in waste-

water, the total ammonium removal capacity will be reduced by high concentrations of other cations. It is therefore important to control processes upstream to minimize other cations, especially calcium. In the case where phosphorus removal by lime precipitation is practiced, this task can be accomplished by optimizing the pH of the recarbonation basins such that calcium solubility is at a minimum.

Ammonium breakthrough. After a clinoptilolite bed has been regenerated, it can be put back on-line and used for ammonium removal again. Initial effluent concentrations of NH_4^+ are typically below 1 mg/L. A typical effluent ammonium curve versus throughput is shown in Figure 6.27.

Regeneration. After the clinoptilolite has reached saturation or the required breakthrough, it must be regenerated. Two methods that have been used to regenerate clinoptilolite are: high pH regeneration using lime, $Ca(OH)_2$, or caustic, NaOH; and neutral pH regeneration using NaCl.

High pH regeneration was the first method developed for regeneration.[75,76] The main advantage of high pH regeneration is the reduced number of bed-volumes necessary to produce an adequate regeneration. Unfortunately, it has one major drawback, precipitation of magnesium hydroxide and calcium carbonate in the ion exchange media. These precipitates are difficult to remove and can cause "mudballing" of the media. Scale and "mudballing" can ultimately reduce the exchange capacity of the media.

Neutral pH regeneration with NaCl has been used to resolve scaling problems.[74] The use of sodium salts in place of other ionic solutions such as Ca increases the amount of ammonium that can be eluted during normal operation of the columns. The main drawbacks of neutral pH regeneration are the increased number of bed-volumes (30 to 40), along with the increased time required for re-

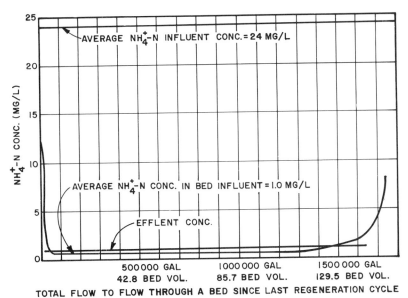

FIGURE 6.27. **Total flow through a bed since last regeneration cycle. NOTE: gal × 3.79 = L.**

generation, and the extra pumping energy required. The main advantage is, of course, the reduction of scaling problems. Figure 6.28 shows ammonia concentrations in the clinoptilolite regenerant solution tanks during neutral pH regeneration.

Regenerant recovery. Regenerant recovery is accomplished by removing the ammonium that has accumulated in the regenerant solution. Once the ammonium has been removed, the regenerant solution can be reused. Two basic methods of regenerant recovery have been developed—gas stripping, and breakpoint chlorination by electrolytic generation of chlorine. Gas stripping can be accomplished by air stripping, steam stripping, or closed gas stream stripping followed by gas absorption.

Design criteria. Design criteria are presented in Tables 6.9 and 6.10 for two plants in the U.S. that have used ion exchange for nitrogen removal. The first is the Rosemount, Minn., treatment plant, which used high pH regeneration and

steam stripping. Its ammonium removal system is no longer in operation because of the high costs associated with the process. (Its nitrogen effluent standards have been relaxed.) The second is the T-TSA Regional Water Reclamation Plant in Truckee, Calif., which uses neutral pH regeneration and closed gas stream stripping with ammonium sulfate recovery. The design criteria presented for T-TSA are the criteria for the new plant expansion, scheduled to be completed in 1984.

Process advantages and disadvantages. Ammonium removal by ion exchange should be considered when both a stringent effluent requirement must be met and climatic conditions cause difficulty with nitrification. Both of these problems are overcome by using ion exchange for ammonium removal. Other advantages to ion exchange are that it produces a relatively low TDS effluent (compared to breakpoint chlorination) and it can produce a product of economic value (fertilizer).

101

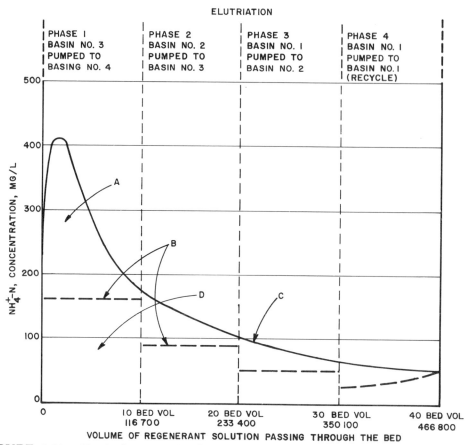

FIGURE 6.28. Typical ammonia elution curve. Regenerant volume in gallons. A. NH$_4^+$-N elutriated current cycle; B. Concentration of NH$_4^+$-N in regenerant solutions going into bed; C. Concentration of NH$_4^+$-N in regenerant solution coming off the bed; and, D. NH$_4^+$-N elutriated in previous cycles. NOTE: gal $\times$ 3.79 = L.

BREAKPOINT CHLORINATION

Chlorine has been used as a disinfectant for well over 100 years. The use of high dosages of chlorine for taste and odor control in water treatment led some operators to observe the increase, disappearance, and subsequent reappearance of chlorine residual as chlorine dosage was incrementally increased. The point at which the transition occurred between chlorine residual disappearance and reappearance came to be known as the "breakpoint." Calvert[77] recognized in 1940 that the breakpoint phenomenon was a result of the oxidation of ammonia nitrogen. Breakpoint chlorination has been studied since that time under controlled laboratory conditions, and it has been applied in certain water and wastewater treatment applications for the removal of ammonia nitrogen from solution.

This chapter provides a summary of the basic chemical equilibria of ammonia nitrogen and chlorine in aqueous solution, and describes the reactions between ammonia and chlorine that leads to the oxidation of ammonia and to end products comprised principally of nitrogen gas. Reaction stoichiometry and reaction rates are also discussed. Data

TABLE 6.9. Design criteria for Rosemount, Minn., final effluent standard.

Final effluent standard	
Parameter	Value
BOD₅, mg/L	10
Suspended solids, mg/L	10
COD, mg/L	10
Ammonium-nitrogen, mg/L	1
Phosphorus, total (P), mg/L	1
pH	8.5
Ion exchange design criteria	
Ammonium exchange columns (two trains of 3)	
Loading rate	4.2 gpm/sq ft, 5.6 BV hr
Clinoptilolite capacity	
per unit volume	0.3 lb/cu ft
per column	90 lb
Ammonium nitrogen loading rate	50 lb/day
Ammonium removal	95%
Clinoptilolite depth per column	6 ft
Clinoptilolite size	20 × 50 mesh
Normal operation	2 columns in series, 250 BV/cycle
Backwash rate	8 gpm/sq ft
Regeneration system	
Brine solution to columns	
Hydraulic application rate	2.0 gpm/sq ft
Volume	4.5 BV
Strength	6% NaCl
Temperature	71°C
pH[c]	11
Brine solution regeneration	
Regeneration cycle length	5 hr
Hydraulic loading rate to steam stripping tower	7 gpm/sq ft
Tower depth	24 ft
Caustic soda added	3 lb/lb NH₄⁺-N
Bed rinse	
Rinse rate	300 gpm
Time	70 min
Ammonia recovery	
Aqueous ammonia strength	1%
Aqueous ammonia volume	1 000 gpd
Ammonia stripper	
Steam @ 10 psig	3 300 lb/hr
Throughput	53 gpm
Size: diameter	3 ft
height	18 ft

are provided from controlled laboratory experiments and full-scale breakpoint applications that show pH and temperature effects on reaction rates and reaction products. Both theoretical and practical discussions are also provided that clarify the breakpoint chlorination process requirement for pH control and, when necessary, alkalinity supplementation. Finally, process control requirements for the feed of both chlorine and the alkalinity supplement are described.

TABLE 6.10.   Design criteria for T-TSA, Calif.

Final effluent standard	
Parameter	**Value**
Suspended solids, mg/L	10
COD, mg/L	45
Total nitrogen, mg/L	9.0
Phosphorus, total, mg/L	0.8
Chlorides, mg/L	200

Ion exchange design criteria	
Parameter	**Value**
Total clino bed flow	6.9 mgd
Ion exchange ammonium removal process	
Number of clino beds	5
Size of clino beds	10' dia $\times$ 40' long
Depth of media	4 ft
Volume each bed	11 670 gal
Media type	Clinoptilolite ion exchange media
Normal bed operation (I)	
Number of beds in service	4
Bed loading rate	6.2 BV/hr
Surface loading rate	3.0 gpm/ft^2
Flow rate oper bed	1.73 mgd
Duration operating cycle	24 hr
Clino bed backwash	
Source of supply	Filter effluent
Backwash rate (max)	8 gpm/ft^2
Backwash flow	1 600 gpm
Ion exchange regeneration system	
Quantity NH$_3$-N per day	1 120 lb/day
Clino bed loading capacity	280 lb NH$_3$-N/bed/cycle
Total BV to exhaustion	148 BV/bed
Bed regeneration	
Duration regeneration cycle	6 hr
Mode of regeneration	1 bed at a time
Regeneration bed loading rate	8.02 BV/hr
Total regeneration BV	40 BV/bed
Regenerant BV transferred (basin to basin)	10 BV
Regeneration flow rate	1 560 gpm
Regenerant supply pumps	3
Pump capacity	1 700 gpm each
Regenerant basins	4 @ 175 000 gal each
Ion exchange regenerant recovery system	
Normal flow rate to process	360 gpm
Regenerant clarifiers	3
Size of clarifiers	20 ft dia
Units in service	2
Normal overflow rate per clarifier	0.57 gpm/ft^2

Final effluent requirements	
Parameter	**Value**
ARRP	
Number of towers	8 (4 strippers and 4 absorbers)
Size of tower	12 ft dia/113 sq ft

TABLE 6.10. Design criteria for T-TSA, Calif. (*continued*)

Final effluent requirements Parameter	Value
Media height	
Stripper tower	9 ft
Absorber tower	3 ft
Gas loading rate	1 450 lb/hr/sq ft = 407 cu ft/min/sq ft
Tower blowers	4 @ 46 000 cfm
Liquid loading rate	
Stripper tower	430 lb/hr/sq ft = 0.82 gpm/sq ft
Absorber tower	520 lb/hr/sq ft = 0.82 gpm/sq ft
Tower supply pumps	2 @ 340 gpm
(NH₄SO₄ storage and transfer system)	
gal/day (NH₄)₂SO₄ @ 40% conc	1 390 gal/day
Sodium chloride system	
Salt used per day (max)	5 340 lb/day
Sulfuric acid system	
Pounds sulfuric acid used/day	4 206 lb/day
Volume 66° baume sulfuric acid used/day	300 gal/day
Sulfuric acid feed pumps	2
Pump capacity	960 gal/day each

Chemistry of chlorine in water. Chlorine gas dissolved in water hydrolyzes rapidly in accord with the expression:

$$Cl_2(aq) + H_2O \rightleftharpoons HOCl + H^+ + Cl^- \quad (21)$$

The hypochlorous acid HOCl is a weak acid that quickly establishes equilibrium with the hypochlorite ion OCl^-, or:

$$K_1 = \frac{[OCl^-][H^+]}{[HOCl]} \quad (22)$$

with the brackets being used to signify molar concentrations. The sum of the chlorine contained in HOCl, OCl^-, and $Cl_2(aq)$ is called "free chlorine," or:

$$c(mg/L \text{ as } Cl_2) = 70.9 \times 10^3 \, ([HOCl] + [OCl^-] + [Cl_2(aq)]) \quad (23)$$

with 70.9 being the gram molecular weight of Cl_2. The quantity of $Cl_2(aq)$ molecules is commonly very small except at a very low pH.

Ignoring the aqueous chlorine, Equations 22 and 23 may be solved simultaneously to obtain the proportion of c in the form of undissociated HOCl, or

$$\frac{[HOCl]}{[c]} = \frac{HOCl \, (mg/L \text{ as } Cl_2)}{c \, (mg/L \text{ as } Cl_2)}$$

$$= \frac{1}{1 + 10^{pH-pK_1}} \quad (24)$$

in which pK_1 is the negative of the common logarithm of K_1. According to Morris:[78]

$$pK_1 = \frac{3000}{T} - 10.068\,6 + 0.0253\,T \quad (25)$$

in which T is the absolute water temperature in K. One-half of c will be HOCl at a pH equal to pK_1 with pK_1 equal to 7.6 at a temperature of 20°C.

Chemistry of ammonia in water. Ammonia gas dissolved in water exists at least in three forms: $NH_3(aq)$, NH_4^+, and NH_4OH. The quantity of NH_4OH molecules as such must be very small[79] and the relation between $NH_3(aq)$ and the ammonium ion NH_4^+ can be expressed for all practical purposes as

$$NH_3(aq) + H^+ \rightleftharpoons NH_4^+ \quad (26)$$

with

$$K_2 = \frac{[NH_4^+]}{[NH_3(aq)][H^+]} \quad (27)$$

The total nitrogen contained in both the $NH_3(aq)$ and NH_4^+ forms of dissolved ammonia gas is:

$$n \, (mg/L \text{ as } N) = 14.0 \times 10^3 \times ([NH_3(aq)] + [NH_4^+]) \quad (28)$$

105

with 14.0 being the gram molecular weight of nitrogen.

Equations 26 and 27 can be solved simultaneously to obtain the proportion of n in the form of $NH_3(aq)$, or:

$$\frac{[NH_3(aq)]}{[n]} = \frac{NH_3(aq) \ (mg/L \ as \ N)}{n \ (mg/L \ as \ N)}$$
$$= \frac{1}{1 + 10^{-pH-pK_1}} \quad (29)$$

According to Bates and Pinching:[80]

$$pK_2 = -\frac{3751.5}{T} + 6.6495 - 0.011032 \ T$$
$$(30)$$

with T being the temperature in K. One-half of n will be $NH_3(aq)$ at a pH equal to $-pK_2$, with pK_2 equal to -9.4 at 20°C.

Formation of chloramines. The free chlorine present at time zero, c_0, represents the weight of Cl_2 added per unit volume of flow. Also, the concentration of ammonia as nitrogen dissolved in the water at time zero is designated by n_0. Finally, the ratio of c_0 to n_0 is called the chlorine:ammonia nitrogen dose application ratio and is designated by the letter M, or $M = c_0/n_0$.

Aqueous ammonia reacts with hypochlorous acid to produce monochloramine, or:

$$NH_3(aq) + HOCl \xrightarrow{k_1} NH_2Cl + H_2O \quad (31)$$

This reaction is slightly reversible, but this can be ignored when the chlorine is present in excess. It also seems to be elementary in the sense that the pH dependency of the reaction rate is correctly predicted as illustrated below.[81]

According to Equation 31, the rate of monochloramine formation is:

$$\frac{d[NH_2Cl]}{dt} = k_1[NH_3(aq)][HOCl]$$

The substitution of Equations 24 and 29 into the above gives:

$$\frac{d[NH_2Cl]}{dt} = \frac{k_1[c][n]}{(1 + 10^{pH-pK_1})(1 + 10^{-pH-pK_2})}$$
$$= K[c][n]$$

in which K is the observed rate of monochloramine formation. The variation of K with pH, pK_1 and pK_2 shown above has been verified experimentally, with the K passing through a maximum with increasing pH. The peak K occurs at a pH of 8.5 at a temperature of 20°C.

Additional reactions which also are often considered to be elementary are:

$$NH_2Cl + HOCl \rightarrow NHCl_2 + H_2O \quad (32)$$

$$NHCl_2 + HOCl \rightleftharpoons NCl_3 + H_2O \quad (33)$$

with the reaction products $NHCl_2$ and NCl_3 being called dichloramine and nitrogen trichloride (trichloramine), respectively. These equations indicate that both $NHCl_2$ and NCl_3 will not be formed readily under alkaline conditions where the OCl^- form of free chlorine predominates. This seems to be the case for Equation 32, but some evidence indicates that NCl_3 may be formed by other reaction pathways at neutral pH values.[82] Both reactions are slightly reversible with the reversibility of Equation 32 commonly being ignored when the free chlorine is present in excess (breakpoint chlorination).

The sum of the chlorine contained in the chloramines is called the combined chlorine residual, or:

Combined c (mg/L as Cl_2) = ([NH_2Cl] + 2[$NHCl_2$] + 3[NCl_3]) 70.9 × 10³

Likewise, the sum of the nitrogen contained in the chloramines is:

Chloramine n (mg/L as N) = ([NH_2Cl] + [$NHCl_2$] + [NCl_3]) 14 × 10³

It should be emphasized that the chloramine nitrogen remains unoxidized. The dissolved ammonia will reappear following decay of the combined residual or when a strong reducing agent is added to the chloramines, as in dechlorination.

Breakpoint reactions. The chlorine-ammonia nitrogen reduction-oxidation (redox) pathways are understood poorly. The idealized reaction scheme described is based on the publications of Morris and Wei.[83,84]

Dichloramine seems to be the key compound as far as redox is concerned. The following reaction has been proposed:

$$NHCl_2 + H_2O \rightleftharpoons NOH + 2\ H^+ + 2Cl^- \quad (34)$$

but it certainly is not elementary.[82] Once formed, the radical NOH may react with mono- or dichloramine to yield nitrogen gas, or:

$$NOH + NH_2Cl \rightarrow N_2 + H^+ + Cl^- + H_2O \quad (35)$$

$$NOH + NHCl_2 \rightarrow N_2 + HOCl + H^+ + Cl^- \quad (36)$$

It may also react with HOCl to produce nitrite, or:

$$NOH + HOCl \rightarrow NO_2^- + 2\ H^+ + Cl^- \quad (37)$$

The free chlorine then will oxidize the nitrite to nitrate.

Equation 36 is interesting in that it indicates that some free chlorine will be recovered as the dichloramine is oxidized. Free chlorine recovery has been observed in the pH region where free chlorine is primarily undissociated HOCl.[82-85]

The nitrogen end-products are primarily N_2 and nitrate. Nitrate production is undesirable because the nutrient remains in a biologically usable form. The trichloramine is a by-product that persists for long periods of time, decomposing back to oxidizable dichloramine very slowly. In other words, the formation of trichloramine is much more rapid than the decomposition of trichloramine in Equation 33.

Reaction stoichiometry. The stoichiometry of the redox reactions depends on the nitrogen end products and the pH of the water containing the reactants. For example, an initial pH sufficiently low to ensure that most of the free chlorine is undissociated HOCl gives:

$$3\ HOCl + 2\ NH_4^+ = N_2 + 3\ H_2O + 5\ H^+ + 3\ Cl^- \quad (38)$$

for an N_2 end-product. Likewise, the stoichiometric equation for a nitrate end-product is:

$$4\ HOCl + NH_4^+ = NO_3^- + H_2O + 6\ H^+ + 4\ Cl^- \quad (39)$$

The acidity produced by chlorine hydrolysis (Equation 21) as well as chlorine-ammonia redox (Equations 38 and 39) can be neutralized by adding an alkalinity supplement before or after the breakpoint reaction has been completed. Nevertheless, a pH drop during redox will always occur with the magnitude of this drop depending on many factors, including the amount and fate of the nitrogen oxidized, the buffering capacity of the water, and the initial water quality characteristics. The maximum stoichiometric yield of protons under neutral or acid conditions is:

$$1.5\ Cl_2(aq) + NH_4^+ = 0.5\ N_2 + 4\ H^+ + 3\ Cl^- \quad (40)$$

and

$$4\ Cl_2(aq) + NH_4^+ + 3\ H_2O = NO_3^- + 10\ H^+ + 8\ Cl^- \quad (41)$$

The reaction stoichiometry also demonstrates that if all of the ammonia nitrogen was oxidized to N_2 then the stoichiometric chlorine:ammonia nitrogen application ratio M would be

$$(1.50)(70.9)/(14.0) = 7.6$$

On the other hand, if only NO_3^- were produced, then the stoichiometric M would be 20.3.

CHARACTERISTICS OF THE BREAKPOINT REACTION

Breakpoint reaction rates. The rate equations for the various reactions listed can be solved simultaneously to yield an idealized breakpoint reaction.[82-85] (Some of the rate constants had to be adjusted empirically for pH dependency, and other factors to fit observed data). To simplify the calculations, the pH was held constant during redox with the results of this modeling being illustrated in Figure 6.29 for an n_0 of 0.50 mg/L and a pH of 7.4.[82] The horizontal axis represents the chlorine:ammonia nitrogen application ratio M, and the vertical axis represents the proportions of n_0 remaining unoxidized at the stated contact time.

According to this model, no nitrogen is oxidized for M ratios of less than 5.0. The proportions of n_0 oxidized, or ending up as NCl_3, increase rapidly with M thereafter. A minimum develops in the ammoniacal nitrogen residual curve because the mono- and dichloramine nitrogen is oxidized much more rapidly than the NCl_3 nitrogen. The residual eventually attains a minimum at the point where the chlorine is just sufficient to oxidize the nitrogen. This point has a M ratio of about 7.8 for the example shown, indicating a small amount of NO_3^- production.

The reliability of Equation 37 for predicting nitrate production is uncertain because this end-product has not been investigated extensively. It is known that M ratios in excess of 9.2 oxidize slightly less than 10% of a 20 mg/L solution of ammonia nitrogen to NO_3^-, with the nitrate production being nearly independent of the pH.[85] The NO_3^- production may decrease as the M ratio is decreased to the exact breakpoint requirement. Pressley et al.[86] reported such a decrease for an n_0 of 20 mg/L with the NO_3^- production varying from 2 to 14% of the total nitrogen oxidized.

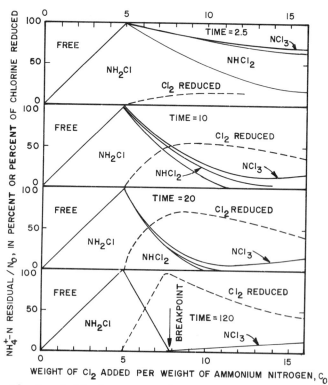

FIGURE 6.29. Idealized chlorine-ammonia reaction in terms of ammonium nitrogen residuals (n_0 = 0.50 mg/L; pH = 7.40; and temperature = 15.0°C).

The effects of pH on the proportion of ammonia nitrogen oxidized in 10 minutes are shown in Figure 6.30 for the case of n_0 equal to 0.50 mg/L, M equal to 8.2, and a temperature of 15°C.[82] The greatest amount of nitrogen was oxidized at a pH of about 7.7 for these particular conditions. Again, it was assumed that the pH remained constant during the oxidation of the nitrogen.

Contact times. The contact time required to oxidize the ammonia nitrogen is of considerable design significance. Such times are not simply stated because the rate of the reaction is dependent, among other factors, on pH, M, and n_0. Figure 6.31 represents an attempt to relate those three parameters with reaction time, which may be used for guidance in very clean water systems.[85]

The M ratio to be used with Figure 6.31 must be equal to or greater than the minimum breakpoint requirement. The vertical axis represents the product of c_0 ($c_0 = M n_0$) and the time t_r required to oxidize at least 99% of the mono- and dichloramine nitrogen. The horizontal axis represents the pH found at time t_r.

The data points shown are restricted to those experiments of Saunier and Selleck[86] having a n_0 of 5 mg/L or less to avoid excessive redox pH drop.

The use of Figure 6.31 may be demonstrated with an example. The water pH at the completion of the reaction is expected to be 7.0. According to the figure, a pH of 7.0 gives:

$$c_0 t_r = M n_0 t_r = 200 \text{ mg/L} \cdot \text{min}$$

The time required to oxidize 99% of all of the nitrogen not contained in NCl_3 is, therefore:

$$t_r(\text{min}) = \frac{200}{M n_0}$$

If M equals 8.0 and n_0 is 0.5 mg/L, then t_r is equal to $(200)/(8 \times 0.5)$, or 50 min. This time would decrease to 5 minutes for a M equal to 8.0 and a n_0 of 5.0 mg/L. The time required to oxidize the nitrogen in NCl_3 would be considerably longer. Clearly, NCl_3 is an undesirable by-product of the breakpoint reaction.

The terminal pH giving the most rapid breakpoint reaction appears to be about 7.5 for the data shown in Figure 6.31. This is slightly less than the 7.7

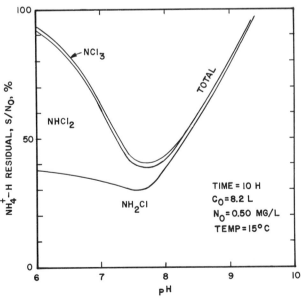

FIGURE 6.30. Effect of reaction pH on the idealized chlorine-ammonium reaction.

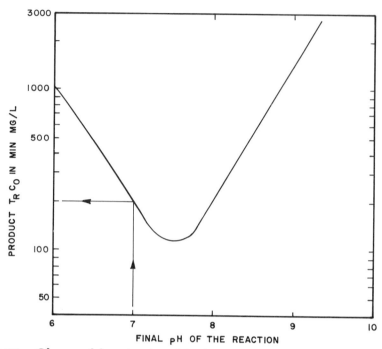

FIGURE 6.31. Observed breakpoint chlorination reaction times. $C_0 = 8.0–13.9:1$; $n_0 = 0.9–5.1$ mg/L; Temp. $= 15–19°C$.

shown in Figure 6.30 for a constant redox pH, but greater than that found to be optimum in full-scale tests.[86] The difference stems at least in part from the redox pH drops that occurred with the observed data. This point will be discussed in more detail later in this chapter.

Mixing. The previous descriptions are based on the assumption of a plug flow reactor (PFR) operation, with the chlorine feed solution being well mixed with the water at time zero. The characteristics of the breakpoint reaction will be altered significantly in a reactor possessing gross backmixing, as demonstrated by the results shown in Figure 6.32 for a completely mixed reactor (a CSTR).[82]

The results shown in Figure 6.32 are for a mean residence time of 32 minutes, a n_0 close to 1.0 mg/L, and a M ranging from 8.9 to 9.7. (Unfortunately the M ratio was not maintained constant from one CSTR experiment to the next, so the curves shown are only approximate.) The results are not directly comparable with those shown in Figure 6.30 for a PFR because the reaction time and n_0 differ, but they do suffice to demonstrate a gross decrease in the rate of nitrogen oxidation. Also, the optimal pH is decreased from about 7.5 to 6.5 or less. These effects probably are insignificant in any contact chamber designed with reasonable care to approximate plug flow characteristics.[82,85]

A means for rapid dispersion of the chlorine solution into the water being treated must be provided, though flash mixers should be avoided on principle because they are small CSTRs. Perforated chlorine solution injector pipes have been found to be adequate in full-scale applications.

Temperature. The breakpoint reaction is temperature dependent, but this factor does not appear to be significant

110

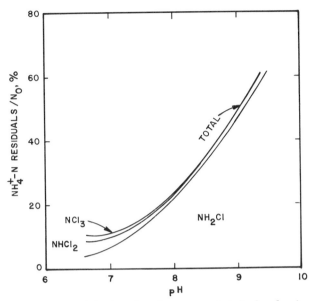

FIGURE 6.32. Breakpoint chlorination in a completely backmixed reactor. NOTE: Detention time-32 min; $C_0 = 8.9\text{-}9.7:1$; $n_0 = 1.0\text{-}1.1$ mg/L; Temp. = 11-17°C.

within the ranges of temperature commonly encountered in municipal wastewaters.[31,84]

BREAKPOINT CHLORINATION IN WASTEWATER

The previous discussions were concerned with reactions of chlorine with ammonia nitrogen only. The characteristics of the breakpoint reaction may change significantly in waters containing other chlorine-demanding substances. For example, the nitrite sometimes found in a partially nitrified activated sludge effluent will exert a rapid chlorine demand. Chlorine also will combine with organic compounds containing ammoniacal nitrogen to form organic chloramines. The organic chloramines then are oxidized at rates differing from the chloramines derived from ammonia. The net effect of all of this is to increase the chlorine dosage requirement and to decrease the apparent speed of the breakpoint reaction.

Breakpoint chlorination is most commonly used to remove the small amounts of ammonia nitrogen remaining after treatment with other nitrogen removal processes. The process is seldom used to remove the relatively high concentrations of ammonia nitrogen found in conventional activated sludge effluents, although this is technically feasible. This was demonstrated in a plant-scale study performed in 1975-76.[86]

The ammonia nitrogen was removed from the partially nitrified effluent of a small activated sludge wastewater treatment plant. The process flow varied from 0.004 to 0.053 m^3/s (0.1 to 1.2 mgd) with the ranges in the pertinent water quality parameters commonly being: 15 to 25 mg/L of ammonia nitrogen (n_0); 1 to 5 mg/L of total organic nitrogen; 0 to 1.4 mg/L of nitrite; and a pH of 7.0 to 7.4. The aqueous chlorine feed solution was neutralized first with sodium hydroxide before being mixed with the water in a small flash mixer. The chlorine contact chamber was a round pipe flowing under pressure with the total mean residence time usually falling within the range of 3 to 5 minutes. The process effluent pH was varied from 6.5

111

to 8.5 with the effluent chlorine residual being maintained at a level of about 12 mg/L, expressed as free chlorine.

The necessary chlorine:ammonia nitrogen application ratio, M, including the excess chlorine required for process control purposes, varied from 8.5 to 12 with a median of 10. The estimated fate of the chlorine applied to the wastewater is shown in Figure 6.33.[86] About 5 to 10% of the chlorine remained unreacted, with 4 to 6% being bound up in NCl_3. The remainder was reduced to chloride or lost to unidentified causes. In terms of the initial ammonia nitrogen concentration, n_0, about 90% of the nitrogen was oxidized to N_2 and 7% to nitrate. Those proportions did not vary significantly for an effluent pH ranging from 6.5 to 8.5.

The oxidation of 15-to-25 mg/L of ammonia nitrogen releases a considerable amount of acidity, which can depress pH severely. The redox pH drop was not measured in this described plant study, but approximate calculations indicated that the pH drops might have been as great as two pH units in some cases (100-fold increase in the H^+ ions).[86] The results of this fieldwork indicated that the speed of the breakpoint

reaction was decreased considerably when the control value for the effluent pH was set much higher than 7.0. For example, a reaction terminating at an effluent pH of 7.5 would commence at a pH of 9.5. The redox reactions should be slow at first (Figure 6.30) until enough acidity is produced to depress the pH. The redox would then speed up as the pH is depressed. The net effect of all of this is to yield the appearance of a much slower breakpoint reaction than predicted under conditions of a (theoretical) constant redox pH.

The effect of redox pH drop in the apparent speed of the breakpoint reaction is illustrated in Figure 6.31 where the results of the wastewater study (dashed line) are compared with those obtained in clean water with much smaller pH drops. The large redox pH drops decreased the apparent speed of the breakpoint reaction by a factor of about three for an effluent pH of 7.5, with the effluent pH giving the most rapid rate of reaction being shifted downward from about 7.5 to 7.0.

In the full-scale field testing, sodium hydroxide was added to the breakpoint process via the chlorine feed solution in amounts sufficient to give a desired pH

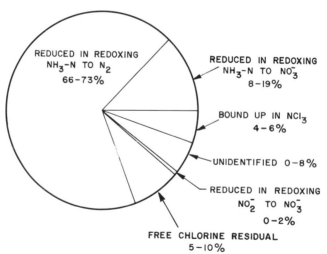

FIGURE 6.33. Fate of chlorine in wastewater breakpoint chlorination ($N_O = 8$ to 22.5 mg/L and effluent pH set point from 7 to 8).

in the process effluent. The pH of the entering water ranged from 7.0 to 7.4, so a return to the same pH would require the addition of the NaOH in approximately stoichiometric amounts as dictated by Equation 40. According to this equation, about 2.7 moles of free hydroxide are required per mole of Cl_2, or 1.5 units of mass of NaOH per unit of mass of Cl_2. The amounts of NaOH actually required ranged from 1.4 to 1.65 units of mass of NaOH per unit of mass of Cl_2 for pH set points of 70 to 7.5, with the median requirement being 1.53 units of mass per unit of mass of Cl_2.

The effluent pH and free chlorine residual were used as control parameters for the application of sodium hydroxide and chlorine, respectively. The free chlorine residual analyzers used in the field demonstration could not absolutely distinguish between a combined and a free residual as illustrated in Figure 6.34. The useful range of the analyzer was thus limited at the low end to values above 3-to-4 mg/L because of the interference from combined residuals extending into the breakpoint region. The minimum reliable process control set point was found to be about 7-to-9 mg/L of free chlorine.

The diurnal flow variation at this plant was too great to permit a simple feedback control on the two chlorina-

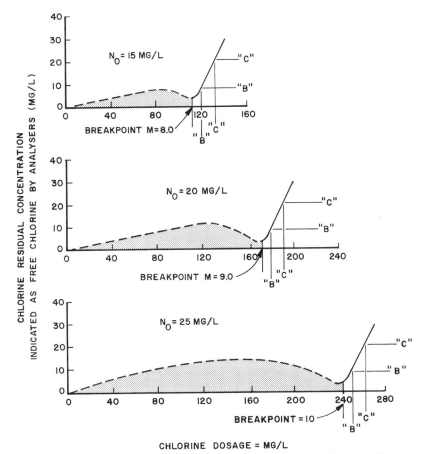

FIGURE 6.34. Chlorine residual analysers—typical interference from combined residuals. (NOTE: "B" control point, 8 mg/L free chlorine residual; "C" 20 mg/L free chlorine residual.)

tors used and the NaOH feed system. The plant flow was monitored continuously, with a flow modified feedback control strategy being used to control the smaller, "trimming" chlorinator and a flow paced control system being used to control the larger, "base load" chlorinator dosage.

The final recommended control systems for this plant are shown in Figures 6.35 and 6.36 for chlorine and alkalinity control, respectively. This represents about the maximum complexity ordinarily required for breakpoint chlorination process control, with the control systems becoming considerably simpler for smaller ammonia nitrogen removals.

Miscellaneous considerations. Breakpoint chlorination increases the TDS in the water. For example, Equation 40 shows that the complete reduction of one mole of Cl_2 produces two moles of chloride ion, that is, one unit of mass of

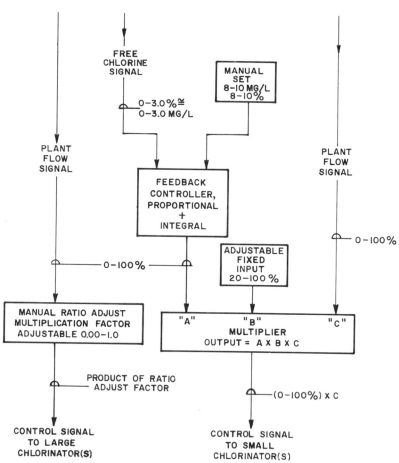

FIGURE 6.35. **Recommended control chlorine feeders. Two chlorinators were used, one large and one small. For this configuration, the large chlorinator was controlled solely by plant flow rate and the small chlorinator was controlled by a flow-modified feedback controller. The control scheme shown had average excursions from the free residual control point of about ±3 mg/L free chlorine. The excursions increased with decreasing plant flow and initial ammonia nitrogen concentrations.**

Cl_2 yields one unit of mass of chloride. The TDS also will be increased with nitrate production as well as by the addition of a base for pH control.

The free chlorine may react with substances contained in wastewater to produce chlorinated organics such as the THM compounds. For example, approximately 90 μg/L of THM (primarily chloroform) were produced in one run of the plant-scale study described previously for an effluent pH set point of 7.0.[87] This production increased to about 300 μg/L at a pH set point of 8.0. The chlorine doses used in the full-scale study were large, but the study results do demonstrate that the potential for THM forma-

tion in that particular activated sludge effluent was considerable.

The breakpoint chlorination process may decrease the dissolved organic nitrogen concentration modestly.[31] It will provide a very good rate of bacterial inactivation.[88] Finally, the concentration of the chlorine residual released to the environment may be restricted by the water pollution control authority. In this case the wastewater may have to be dechlorinated.[89]

SUMMARY OF TREATMENT ALTERNATIVES

Nitrogen removal from wastewater can be achieved by either biological or

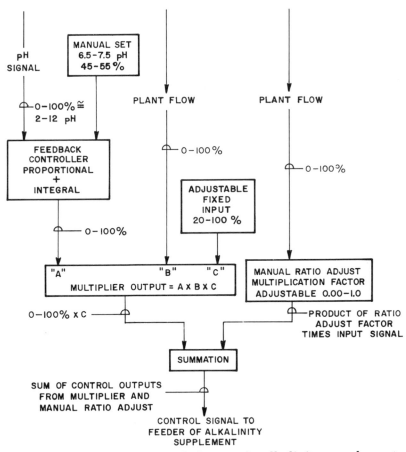

FIGURE 6.36. Recommended control—alkalinity supplement.

physical/chemical means. Selection of the appropriate process depends on several factors:

- pH,
- temperature,
- type and operational modes of preceding processes,
- the nature of recycle streams,
- effluent restrictions, and
- cost

Biological denitrification. Biological denitrification is the most widely used process in wastewater treatment where nitrogen removal is required. In this process, facultative, heterotrophic bacteria reduce nitrate in a two-stage process in which nitrite is the intermediate product and the final product is nitric oxide, nitrous oxide, or nitrogen gas, with nitrogen gas predominating. Denitrifying bacteria also can convert nitrite to ammonia for cell synthesis, but this normally will not occur if sufficient ammonia is present to meet growth requirements.

Denitrifying bacteria work best at pH between 7 and 8 and temperature between 0° and 30°C. It is not ordinarily necessary to control these variables because denitrifying wastewater treatment plants commonly operate within these ranges. More important is the presence of organic substrates in the wastewater. Often methanol or other organic substrates must be added so that a slight excess of organic carbon will exist to ensure denitrification rates close to maximum.

Suspended growth reactors. Suspended growth reactors can be operated as either separate or single-sludge systems in either the pre-denitrification or post-denitrification mode. Single-sludge, pre-nitrification systems for combined carbon oxidation, nitrification, and denitrification are the most common and usually the most cost effective. Design is based on sludge age, total volatile solids, or the unit denitrification approach.

Historically, the unit denitrification rate approach has been used most often. Bench-scale testing is usually necessary to establish design criteria.

Another important design consideration is the organic carbon source for the denitrification reaction. Although methanol has been used extensively in the past, the cost of petrochemical products has made alternate sources such as wastewater and waste carbon streams increasingly attractive.

Fixed-film reactors. Fixed-film reactors include biological fluidized beds, packed (fixed) beds of either highly porous media or low porosity fine media, nitrogen gas-filled packed beds, and rotating biological contactors. As is the case for suspended growth systems, care must be taken to exclude oxygen from the reactor. Excluding oxygen is normally accomplished by completely submerging the reactor media in the wastewater.

Physical/chemical denitrification

Ammonia stripping. In the ammonia-stripping process, ammonium nitrogen in the secondary effluent is converted from ammonium ion (NH_4^+) to a gas by raising the pH. The wastewater is then pumped to a tower to be contacted with air to strip the ammonia from solution. The tower may or may not (in the case of a spray tower) have packing, with a fan at either the top (induced draft) or base (forced draft, countercurrent). This process is found most commonly at plants using lime treatment for phosphorus removal because the lime-treated and clarified wastewater is at or near the optimum pH for air stripping. It may also be practical if ammonia concentrations are so high that they are toxic to biological treatment processes.

The advantages of ammonia stripping are that it is a relatively simple process that does not create any residual solids requiring disposal and that it has no significant recycle flow streams that affect other processes. A major process limitation is the effect of temperature on removal efficiency—as the air temperature drops, the efficiency of the process also drops. The tower becomes inoperable when the air temperature falls very far below freezing. It is not usually practical to heat the large volume of air required for the stripping process or to heat the liquid, unless waste heat from another process is available. Some installations have also experienced a problem with calcium carbonate scale depositing on the tower packing and structural members.

The principle factors affecting design and performance of ammonia-stripping towers include the following:

- tower configuration,
- pH,
- temperature,
- hydraulic loading,
- tower packing depth and spacing,
- air flow, and
- scale control.

Typical design criteria for ammonia stripping towers are:

- Hydraulic loading—40 to 80 L m^2· min (1 to 2 gpm/sq ft),
- air/water ratio—2 240 to 3 700 m^3/m^3 (300 to 500 cu ft/gal),
- air pressure drop—0.12 to 0.31 kPa (0.5 to 1.25 in. water),
- fan tip speed—2 700-3 700 m^3/min (9 000 to 12 000 fpm),
- packing depth—6 to 7.6 m (20 to 25 ft),
- packing spacing—50 to 100 mm (2 to 4 in.) horizontal and vertical, and
- packing type—wood or plastic.

Breakpoint chlorination. When chlorine is added to wastewater containing ammonium nitrogen, it will react with the nitrogen to produce chloramines. With further addition of chlorine to the breakpoint of reaction, the chloramines will be converted primarily to nitrogen gas. Chlorine gas and sodium hypochlorite are the most common chlorine sources used.

The reaction rate and the optimum chlorine dose are dependent on the degree of treatment the wastewater has received, pH, and the type of mixing provided. Although the breakpoint reaction is temperature dependent, it is not affected in the temperature range commonly encountered in municipal wastewater.

The breakpoint chlorination reaction produces an equivalent unit of mass of hydrochloric acid, which will depress the pH of the wastewater if insufficient natural alkalinity is present. If the pH is too low, intolerable byproducts such as dichloramine and odorous nitrogen trichloride can form and the wastewater may become unacceptable for discharge. pH can be adjusted by adding a base such as lime or sodium hydroxide to the chlorine feed.

The principal advantages of the breakpoint chlorination process are low capital costs, assurance of a high degree of disinfection, and a 90 to 95% conversion of ammonium to nitrogen gas, which can be dissipated harmlessly to the atmosphere. Furthermore, the breakpoint process is comparatively reliable with the proper control of pH and chlorine dose.

Some of the disadvantages of breakpoint chlorination are high operating costs, increased total dissolved solids (particularly a problem if the effluent becomes a drinking water source), complexity of control, and possible formation of undesirable byproducts, especially THM compounds.

Design considerations for breakpoint chlorination are as follows:

- Mixing, as a means for rapid dispersion of the chlorine solution into the water treated, must be provided, although flash mixers should be avoided;
- contact time required to oxidize the ammonia nitrogen is dependent on pH, the stoichiometric chlorine : ammonia nitrogen application ratio, the initial ammonia concentration, and the presence of other chlorine demanding substances;

• chlorine dosage as part of the chlorine : ammonia dose application ratio is suggested to be maintained at 10 : 1; and

• facilities must be provided for feeding an acid neutralizing compound to maintain the wastewater pH within the range yielding the desirable breakpoint reaction.

Selective ion exchange. The selective ion exchange process derives its name from the use of zeolites that are selective for NH_4^+ relative to calcium, magnesium, and sodium. The zeolite currently favored for this use is clinoptilolite,

which occurs naturally in several extensive deposits in the western U.S.

The NH_4^+ ion is removed by passing the wastewater through a bed of clinoptilolite at a rate of about 6 to 10 bed volumes per hour. Removal of 96% may be expected with influent NH_4^+-N concentrations of about 20 mg/L.

The capacity of the clinoptilolite is nearly constant over the pH range of 4 to 8, but diminishes rapidly outside this range. Wastewater composition has an effect on the exchange capacity. For relatively constant influent NH_4^+-N concentrations, the exchange capacity de-

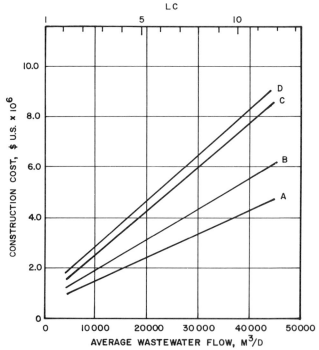

FIGURE 6.37. Construction cost for carbon and nitrogen removal portion of a new plant. Cost Basis: ENR Index 3000; Curves prepared from separate cost estimates for three plant sizes located at Penticton, British Columbia; and effluent objectives are:

Total BOD$_5$	20 mg/L
Suspended Solids	20 mg/L
NH$_4$-N	1.0 mg/L
NO$_3$ + NO$_2$-N	5.0 mg/L
Total P	0.6 mg/L
pH	6.6

creases sharply with increasing concentrations of competing cations, such as calcium and magnesium. Ammonium removal to residual levels below 0.5 mg/L is technically feasible, but requires shorter service cycles and greater regeneration requirements. Flow rates in the range of 7.5 to 15 BV/hr indicate no effect on ammonium effluent values.

Typical design criteria for the clinoptilolite beds are as follows:

- Volumetric loading: 6 to 10 BV/hr;
- surface loading: 160 to 240 L/m²·min (4 to 6 gpm/sq ft);
- bed volumes to exhaustion: 150 to 200;

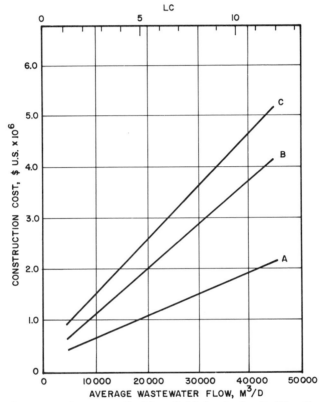

FIGURE 6.38. Construction cost "add-on" biological nitrification/denitrification facilities at an existing plant. Cost Basis: ENR Index 3000; Curves prepared from separate cost estimates for three plant sizes located at Penticton, British Columbia; and effluent objectives are:

Total BOD₅	20 mg/L
Suspended Solids	20 mg/L
NH₄-N	1.0 mg/L
NO₃ + NO₂-N	5.0 mg/L
Total P	0.6 mg/L
pH	6.5

Costs assume no major renovations required to existing final clarifiers.

• backwash rates: 240 to 330 L/m^2·min (6 to 8 gpm/sq ft);
• depth of media: 1.8 m (6 ft); and
• influent calcium concentration: 40 to 50 mg/L.

COSTS OF NITROGEN REMOVAL

Processes for the removal of nitrogen vary widely in construction costs as well as operation and maintenance cost. In addition, power requirements and use of chemicals also vary from process to process. Each method will therefore exhibit different sensitivity to changes in power, labor, and chemical costs.

Capital costs. From the standpoint of capital costs, accurate projections are difficult because a limited number of installations exist today and these are generally small in size and designed for specialized situations. Accurately projecting capital costs is further complicated by the fact that such data as exists does not correlate well between

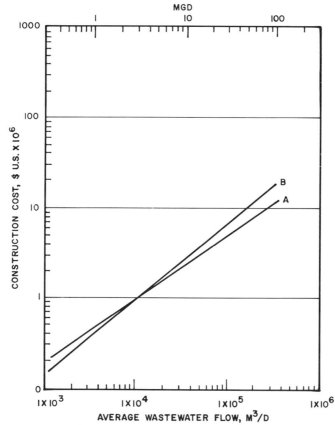

FIGURE 6.39. Construction cost for ammonia removal by high pH ammonia stripping. Cost Basis: ENR Index = 3000; does not include lime clarification or sludge handling and disposal; design objectives are:

Air Flow	2960 m^3/m^3
Pump TDH	150 kPa
Influent NH$_4$-N	15 mg/L
Effluent NH$_4$-N	2–3 mg/L

sources. This is partially because of varying design assumptions and different cost bases. But most importantly, it is because sources have included or excluded varying cost elements such as pumped versus gravity flow, covered versus uncovered basins, land, electrical, and instrumentation and control loops.

Figures 6.37 through 6.41 present capital cost data from a number of sources for nitrogen removal by biological nitrification/denitrification, ammonia stripping, breakpoint chlorination, and selec-

tive ion exchange. In all cases, the information presented has been adjusted from the basis used in the original studies[90-93] to a uniform ENR 3 000 index and an attempt has been made to standardize contingency, legal, fiscal, engineering, and construction interest costs at 35%. Land costs are excluded.

The design engineer may prefer to make project-specific estimates rather than rely on the paucity of cost data available. In general, existing cost curves appear to be low, perhaps by 30% or more. The problem of cost data

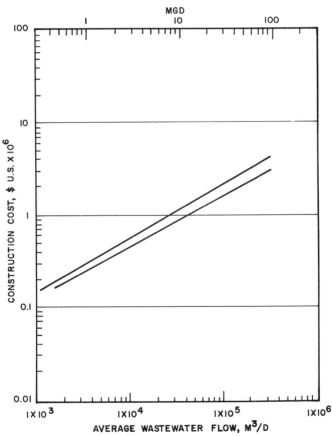

FIGURE 6.40. **Two construction cost curves for ammonium removal by breakpoint chlorination. Cost Basis: ENR Index = 3000; Gas/Solution Cl$_2$ feed; uncovered tanks, no off gas stripping; computer control not included; no dechlorination; and, contact time of .30 minutes.**

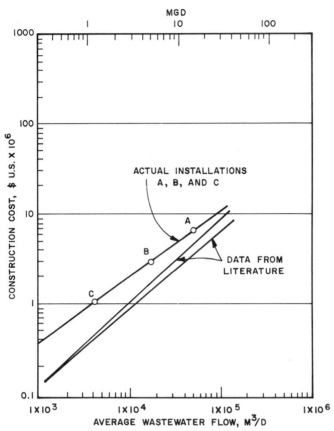

FIGURE 6.41. Construction cost for ammonium removal by selective ion exchange with regenerant renewal by closed cycle stripping. Cost Basis: ENR Index = 3000.

extrapolation is illustrated by the relatively few actual data points plotted on the figures.

Operation and maintenance costs. Figures 6.42 and 6.43 present operating and maintenance (O&M) costs for new and "add-on" biological nitrification/denitrification facilities. These curves were based on the unit costs for labor, power, and chemicals at Pentiction, British Columbia, as presented in Table 6.11. It should be noted that these unit costs vary slightly from those used in estimating O&M costs for physical/chemical nitrogen removal alternatives. Because curves were not presented for the various components of the O&M cost, it was not possible to adjust the biological nitrification/denitrification curves to the same basis used in preparing the physical chemical nitrogen removal O&M cost estimates.

Figures 6.44 through 6.46 present O&M

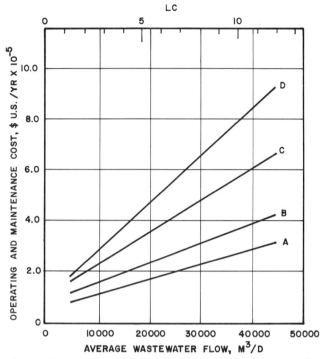

FIGURE 6.42. Operating and maintenance cost for carbon and nitrogen removal portion of a new plant.

cost data for nitrogen removal by ammonia stripping, breakpoint chlorination, and selective ion exchange. These data are taken directly from a 1978 EPA publication.[91] The basis for costs used is listed in Table 6.12. Here also, the design engineer may find these cost data to be low. In particular, power usage figures shown are, for at least some unit processes, radically lower than actual operating experience would dictate. For this reason, Figures 6.47 through 6.50, taken from another EPA publication,[94] are included. These portray annual electrical energy use as a function of capacity for the same unit processes. These data may be more realistic than those displayed in Figures 6.44 through 6.46.

TABLE 6.11. Basis for estimating operating and maintenance costs for biological nitrification/denitrification facilities.

Item	Unit cost
Labor	$22 000/man-year
Power	$0.01/kWh
Lime	$72.70/mg
Methanol	$0.22/L
Pickle liquor	$2.11/100 kg
Chlorine	$319/mg
Polymer	$5.29/kg
Equipment replacement and supply	0.6% total capital cost/yr

Note: The cost of labor and chemicals may vary widely between areas and are constantly changing. The above values represent costs in Penticton, B.C., at a particular point in time.

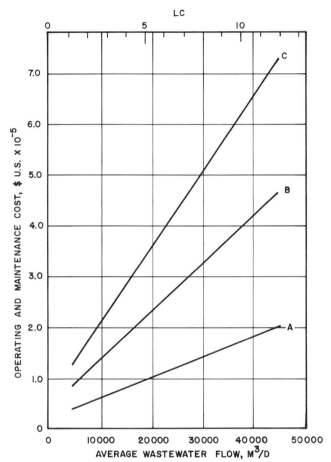

FIGURE 6.43. Operating and maintenance cost for "add-on" biological nitrification/denitrification facilities at an existing plant.

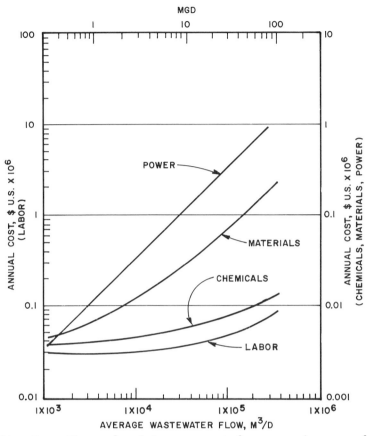

FIGURE 6.44. Operation and maintenance costs for ammonia removal by high pH stripping.

Cost Basis:
Power	$0.02/kWh
Labor	$7.50/man hr
Pump TDH	150 kPa
Pump Efficiency	60%
Hydraulic Load	40 L/m$^2 \cdot$min
Air Flow	2960 m^3/m^3
Influent NH$_4^+$-N	18 mg/L
Effluent NH$_4$-N	3 mg/L

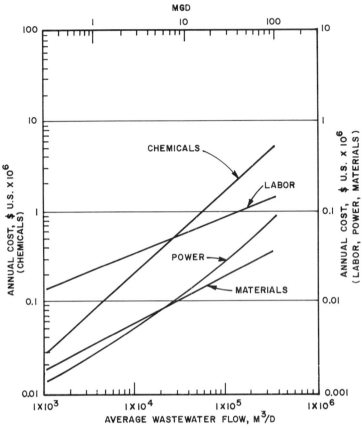

FIGURE 6.45.   Operation and maintenance costs for breakpoint chlorination.

Cost Basis:

Power	$0.02/kWh
Labor	$7.50/man hr
Lime	$160/ton
NH_4^+-N Influent	19 mg/L
NH_4^+-N Effluent	2 mg/L
Cl_2 Dose	10 mg/mg NH_4^+-N
Lime Dose	0.9 mg/mg Cl_2

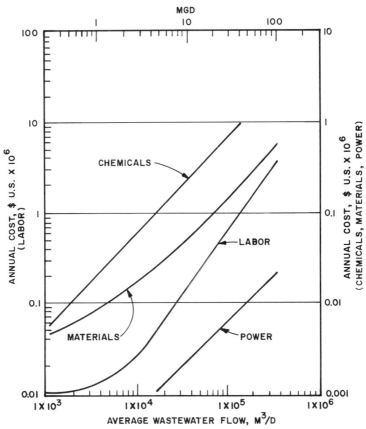

FIGURE 6.46.   Operation and maintenance costs for ammonia removal by selective ion exchange with regenerant renewal by closed cycle stripping.

Cost Basis:

Power	$0.02/kWh
Labor	$7.50/man hr
Sulfuric Acid	66° Be $50/ton
Regeneration	40 Bed Volumes/Cycle
Throughput	100–150 Bed Volumes/Cycle
Regenerant	2% NaCl Solution
Salt Use	0.01 kg/m^3 throughput
Caustic Soda	0.9 mg NaOH/mg NH$_4^+$-N removed

TABLE 6.12. **General cost basis for operation and maintenance costs for physical/chemical nitrogen removal facilities.**

Energy costs	
A. Electric power	= $0.02/kWh
B. Fuel oil	= $0.37/gal
C. Gasoline	= $0.60/gal
Land	= $1,000/acre
Chemical costs	
A. Liquid oxygen	= $65/ton
B. Methanol	= $0.50/gal
C. Chlorine 150 lb. cylinder	= $360/gal
1 ton cylinder	= $260/ton
Tank car	= $160/ton
D. Quicklime	= $25/ton
E. Hydrated lime	= $30/ton (as CaO)
F. Polymer (dry)	= $1.50/lb.
G. Ferric chloride	= $100/ton
H. Alum	= $72/ton
I. Activated carbon (granulated)	= $0.50/lb.
J. Sulfuric acid (66° Be)	= $50/ton
K. Sodium hexametaphosphate	= $0.25/lb.
L. SO_2 150 lb. cylinders	= $450/ton
1 ton cylinder	= $215/ton
Tank car	= $155/ton

Basis of Costs (All units U.S.)

Labour rate, including fringe benefits = $7.50/hr.

Note: Labor costs are based on a man-year of 1500 hours. This represents: a 5-day work week; an average of 29 days for holidays, vacations, and sick leave; and 6½ hours of productive work time per day.

Design basis

Operation and maintenance costs are based on design average flow.

Operation and maintenance costs include:

A. Labour costs for operation, preventive maintenance, and minor repairs.
B. Materials costs to include replacement parts and major repair work (normally performed by outside contractors).
C. Chemical costs.
D. Fuel costs.
E. Electrical power costs.

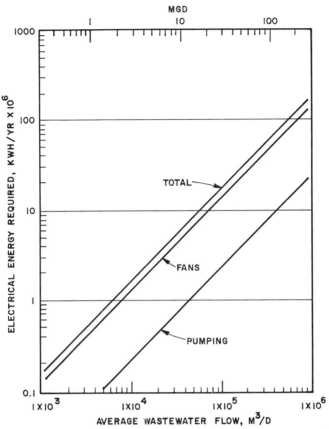

FIGURE 6.47. Primary energy requirements for high-pH ammonia stripping.

Design Assumptions:

Influent pH	11
Air Temperature	21°C
Influent NH_4^+-N	15 mg/L
Stripping Efficiency	80%
Hydraulic Loading	40.75 L/m²·min
Air/Water Ratio	0.3 m³/L

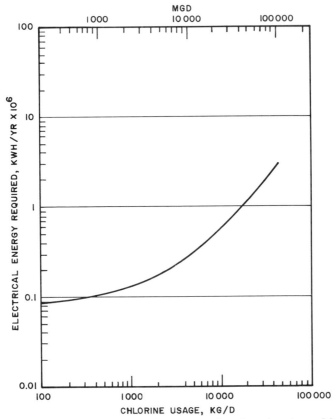

FIGURE 6.48. Primary energy requirements for breakpoint chlorination with sulphur dioxide dechlorination.

Design Assumptions:

Influent NH_4^+-N	15 mg/L
Effluent NH_4^+-N	0.1 mg/L
Dosage Ratio Cl_2 NH_4^+-N	8:1
Residual Cl_2	3 mg/L
Detention in Rapid Mix	1 min
Sulfur Dioxide Feed	
Ratio, SO_2:Cl_2	1:1

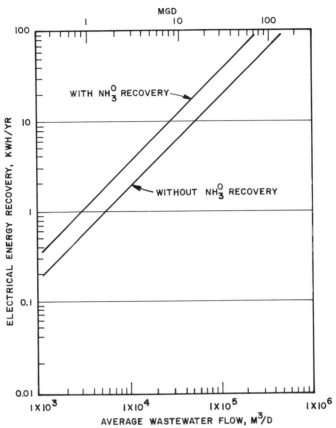

FIGURE 6.49. Primary energy requirements for ammonium removal by ion exchange-regenerant renewal by air stripping.

Design Assumptions: Regenerant softened with NaOH and Clarified at 33 $m^3/day \cdot m^2$.

Influent to Clino. Bed, NH_4^+-N	15 mg/L
Throughput	150 Bed Volumes/Regeneration
Regeneration	40 Bed Volumes Regeneration/Cycle
Regenerant Stripper	31 m^3/m^2
Air Flow	4.23 m^3 air/L
Stripper Height	10 m

NH$_3$ recovery by absorber tower with H$_2$SO$_4$

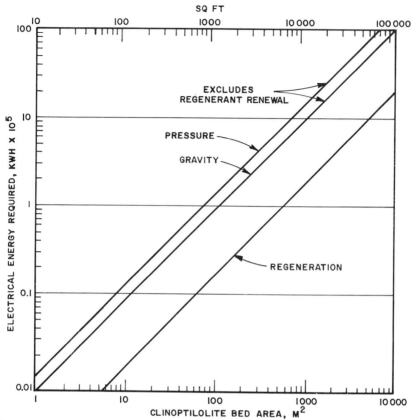

FIGURE 6.50. Primary energy requirements for ammonium removal by ion exchange with regeneration and excluding regenerant renewal.

Design Assumptions:

Influent NH_4^+-N	15 mg/L
Influent SS	less than or equal to 5.0 mg/L
Throughput	150 Bed Volumes
Loading Rate	6 Bed Volumes/hr
Head Avail. to Gravity	22 kPa
Pump Head for Pressure	30 kPa
Regeneration	40 Bed Volumes/Cycle
Regeneration Timing	1 Regeneration/day
Regeneration Pump Head	30 kPa
Bed Depth	1.2 m

REFERENCES

1. Sawyer, C. N., and Bradney, L., "Rising of Activated Sludge in Final Settling Tanks." *Sew. Works J.*, **17**, 6, 1191 (1945).

2. Brandon, T. H., and Grindley, J., "Effects of Nitrates on the Rising of Sludge in Sedimentation Tanks." *J. Proc. Inst. Sew. Purif.*, 175 (1944).

3. Davies, T., "Population Description of a Denitrifying Microbial System." *Water Res.*, **5**, 553 (1971).

4. Delwiche, C. C., "Denitrification." "A Symposium on Inorganic Nitrogen Metabolism." McElroy and Glass (Eds.), John Hopkins Press, 233 (1956).

5. Sperl, H., "Denitrification with Methanol: A Selective Enrichment for Hyphomicrobium Species." *J. Bacteriol.*, **108**, 733 (1971).

6. Fewson, N., "Nitrate reductase from Pseudomonas Aeruginosa." *Bichim. Biophys. Acta*, **49**, 335 (1961).

7. Painter, "A Review of Literature on Inorganic Nitrogen Metabolism in Micro-Organisms." *Water Res.*, **4**, 393-450 (1970).

8. Nason, T., "Inorganic Nitrogen Metabolism." *Ann. Rev. Microbiol.*, **1**, 203 (1958).

9. Nason, T., "Enzymatic Pathways of Nitrate, Nitrite and Hydroxylamine Metabolisms." *Microbiol. Rev.*, **26**, 16 (1962).

10. Delwiche, C. C., "The Nitrogen Cycle." *Sci. Am.*, 223, No. 3, pp. 137-146 (1970).

11. Myers, J., and Natsen, F. A., "Kinetic Characteristics of Warburg Manometry." *Arch. Biochem. & Biophys.*, **55**, 373 (1955).

12. Schmidt, B., and Kampf, W. D., "Uber den Einfluss des Saverstoffe auf die Denitrifikationsleistung von *Pseudomonas fluorescens, Arch. Hrg. Bakt.*, **146**, 171 (1962).

13. Skerman, V. B. D., and Macrae, I. C., "The Influence of Oxygen on the Reduction of Nitrate by Adapted Cells of Pseudomonas Denitrificans." *Can. J. Microbiol.*, 3215 (1957).

14. Wuhrmann, "Effects of Oxygen Tension on Biochemical Reactions in Sewage Purification Plants." In "Advances in Biological Waste Treatment." Proc. 3rd Conf. Biol. Waste Treat., Munich, 27 (1960).

15. Christensen, M. H., and Harremoes, P., "Biological Denitrification in Wastewater Treatment." Rep. 2-72, Dep. of Sanit. Eng., Tech. Univ. of Denmark, 51 (1972).

16. Krul, J. M., "The Relationship between Dissimilatory Nitrate Reduction and Oxygen Uptake by Cells of an Alcaligenes Strain in Flocs and in Suspension and by Activated Sludge Flocs." *Water Res.*, **10** (1976).

17. Pasveer, A., "Beitrag uber Stickstoffbeseitigung aus Abwassern." (Contribution on Nitrogen Removal from Sewage), Munchner Beitrage zur Abwasser-Fisherei—und Flussbiologie, Liebermann (Ed.) **Bd. 12**, 197 (1965).

18. Climenhage, D. C., and Stelzig, A., "Biological Process for Nitrogen—BOD Removal at Maitland Works, Du Pont of Canada." 20th Ontario Ind. Waste Conf. (1973).

19. McCarty, P. L., et al., "Biological Denitrification of Wastewaters by Addition of Organic materials." In Proc. 24th Ind. Waste Conf., Lafayette, Indiana: Purdue University (1969).

20. Jeris, J. S., and Owens, R. W., "Pilot Scale High Rate Biological Denitrification at Nassau County, N.Y." Pres. Winter Meeting of New York Water Pollut. Control Assoc. (Jan. 1974).

21. Horstkotte, G. A., et al., "Full-scale testing of a water reclamation system." *J. Water Pollut. Control Fed.*, **45**, 181 (1974).

22. Delwiche, ., "Denitrification." In "A Symposium of Inorganic Nitrogen Metabolism." McElroy and Glass (Eds.), John Hopkins Press (1956).

23. Dawson, R. N., and Murphy, K. L., "Factors Affecting Biological Denitrification in Wastewater." In "Advances in Water Pollution Research." S. H. Jenkins (Ed.), Pergamon Press, Oxford, England, (1973).

24. Clayfield, G. W., "Respiration and Denitrification Studies on Laboratory and Works Activated Sludges." *Water Pollut. Control*, London, **73**, 51 (1974).

25. Christensen, M. H., and Harremoes, P., "Biological Denitrification of Sewage: A Literature Review." *Prog. in Water Technol.*, **8**, 509 (1977).

26. Sutton, P. M., et al., "Continuous Biological Denitrification of Wastewater Technology Development Report." EPS 4-WP-74-6 Environ. Prot. Serv., Environment Canada, 20 (1974).

27. McCarty, P. L., "Nitrification-Denitrification by Biological Treatment." Correspondence Conf. on Denitrification of Munic. Wastes. Univ. of Mass. (1973).

28. Dawson, R. N., "Batch Studies on the Biological Denitrification of Wastewater." Ph.D. thesis, McMaster Univ. (1971).

29. Stensel, H. D., "Biological Kinetics of the Suspended Growth Denitrification Process." Ph.D. thesis, Cornell Univ. (1971).

30. Mulbarger, M. C., "Nitrification and denitrification in activated sludge systems." *J. Water Pollut. Control Fed.*, **43**, 2040 (1971).

31. "Process Design Manual for Nitrogen Control." U.S. EPA, Technology Transfer, Washington, D.C. (1975).

32. Engberg, D. J., and Schroeder, E. D., "Kinetics and Stoichiometry of Bacterial Denitrification as a Function of Cell Residence Time." Univ. of Calif. at Davis, Unpbl. paper (1974).

33. Moore, S. F., and Schroeder, E. D., "The Effect of Nitrate Feed Rate on Denitrification." *Water Res.*, **5**, 445 (1971).

34. Requa, D. A., "Kinetics of Packed Bed Denitrification." Thesis submitted in partial satisfaction of the requirements for the degree of Master of Science in Engineering, Univ. of Calif. at Davis (1970).

35. Requa, D. A., and Schroeder, E. D., "Kinetics of packed bed denitrification." *J. Water Pollut. Control Fed.*, **45**, 1696 (1973).

36. Parker, D. S., "Case Histories of Nitrification and Denitrification Facilities." Prepared for the EPA Technol. Transfer Program, (May 1974).

37. Parker, D. S., et al., "Development and Implementation of Biological Denitrification for Two Large Plants." Presented at the Conf. on Nitrogen as a Water Pollutant, sponsored by the IAWPR, Copenhagen, Denmark, (August 1975).

38. "Denitrification for Anaerobic Filters and Ponds." Phase II. Kerr, R., Water Research Center, EPA WPCRS 13030 ELY 06/71-14, (June 1971).

39. Denitrification for Anaerobic Filters and Ponds. Kerr. R., Water Research Center, EPA WPCRS 13030 ELY 04171-8, (April 1971).

40. Murphy, L. K., and Sutton, P. M., "Pilot Scale Studies on Biological Denitrification." Presented at the 7th International Conference on Water Pollution Research. Paris (September 1974).

41. Requa, D. A., and Schroeder, E. D., "Kinetics of Packed Bed Denitrification." *J. Water Pollut. Control Fed.*, **45**, 1696 (1973).

42. Sutton, P. N., et al., "Low Temperature Biological Denitrification of Wastewater," *J. Water Pollut. Control Fed.*, **47**, 122 (1975).

43. "Description of the El Lago, Texas Advanced Wastewater Treatment Plant." Seabrook, Texas: Harris County Water Control and Improvement District Number 50 (March 1974).

44. Ecolotrol, Inc., "Hy-Flo Fluidized Bed Denitrification." *Ecolotrol Tech. Bull.*, No. 123-A (November 1974).

45. Jeris, J. S., and Owens, R. W., "Pilot Scale High Rate Biological Denitrification." *J. Water Pollut. Control Fed.*, **47**, 2043 (1975).

46. Jeris, J. S., and Owens, R. W., "Biological Fluidized Beds for Nitrogen Control." In "Advances in Water and Wastewater Treatment-Biological Nutrient Removal," M. Wanichsita and W. Eckenfelder (Eds.), Ann Arbor Science (1978).

47. Francis, C. W., and Malone, C. D., "Anaerobic columnar denitrification of high-nitrate wastewater." *Progr. Water Technol.*, **8**, 1977, (4/5), 687–711.

48. Antonie, R. L., "Nitrogen Control with the Rotating Biological Contactor." In "Advances in Water and Wastewater Treatment-Biological Nutrient Removal" M. Wanichsita and W. Eckenfelder (Eds.), Ann Arbor Science Pub. Inc. (1978).

49. Sutton, P. M., et al., "Single Sludge Nitrogen Removal Systems." Res. Rep. No. 88, Canada-Ontario Agreement on Great Lakes Water Quality (1979).

50. Sutton, P. M., et al., "Nitrification and Denitrification of an Industrial Wastewater: Research and Demonstration Studies." Presented at 51st Annual WPCF Conf., Anaheim, Calif. (1978).

51. Bridle, T. R., et al., "biological Nitrogen Control of Coke Plant Wastewaters." *Water Sci. and Technol.*, **13**(1), 667 (1981).

52. Dawson, R. N., and Murphy, K. L., "Factors Affecting Biological Denitrification of Wastewater." In "Advances in Water Pollution Research." Jerusalem 671–683 (1972).

53. Sutton, P. M., "Continuous Biological Denitrification of Wastewater." Master thesis, McMaster Univ. (1973).

54. Climenhage, D. C., "Biological Denitrification of Nylon Intermediate Wastewater." Presented at 2nd Canadian Chem. Eng. Conf. (September 1972).

55. Wilson, T. E., and Newton, D., "Brewery Wastes as a Carbon Source for Denitrification at Tampa, Florida." Proc. 28th Ann. Purdue Ind. Waste Conf., Purdue, Indiana (May 1973).

56. Monteith, H. D., et al., "Industrial Waste Carbon Sources for Biological Denitrification." Presented at the 10th IAWPR Conf., Toronto (1980).

57. McCarty, P. L., et al., "Biological Denitrification of Wastewaters by Addition of Organic Materials." Proc. 24th Ind. Waste Conf., Purdue, Indiana (1973).

58. Stern, L. B., and Marais, G. V. R., "Sewage as an Electron Donor in Biological Denitrification." Res. Rep. No. W7, Univ. of Cape Town, South Africa (1974).

59. Marsden, M., and Marais, G. V. R., "Role of the Primary Anoxic Reactor in Denitrification and Biological Phosphorus Removal." Res. Rep. No. W19, Univ. of Cape Town, South Africa (1976).

60. Sun-nan Hong, et al., "A Biological Wastewater Treatment System for Nutrient Removal." Presented at 54th WPCF Conf., Detroit, Mich. (1981).

61. Schwinn, D. E., and Slorrier, D. F., "One Step for Nitrogen Removal—A Giant Step for Optimum Performance." *Water & Wastes Eng.*, 33–37 (Dec. 1978).

62. Bridle, T. R., et al., "Start-up of Full Scale Nitrification-Denitrification Treatment Plant for Industrial Waste." Proc. 31st Purdue Ind. Waste Conf., Purdue Univ., Indiana (1976).

63. Bridle, T. R., et al., "Operation of a Full-Scale Nitrification-Denitrification Industrial Waste Treatment Plant." *J. Water Pollut. Control Fed.*, **51**, 127 (1979).

64. Wilson, R. W., et al., "Design and Cost Comparison of Biological Nitrogen Removal Systems." Presented at 51st Annual WPCF Conf., Anaheim, Calif. (1978).

65. Culp, R. L., and Culp, G. L., "Advanced Wastewater Treatment." VanNostrand Reinhold, New York (1971).

66. Tchobanoglous, G., "Physical and Chemical

Processes for Nitrogen Removal: Theory and Applications." Presented at 12th Sanit. Eng. Conf., Univ. of Ill. (Feb. 1970).

67. U.S. Environ. Prot. Agency, "Cost-Effective Wastewater Treatment Systems." Tech. Report EPA-430/9-75-002, Washington, D.C. (July 1975).

68. Prather, B. V., and Gandy, A. F., Paper presented at 29th Midyear Meeting of the American Petroleum Institute's Division of Refining, St. Louis, Mo. (May 1964).

69. Prather, B. V., Paper presented at 13th Industrial Waste Conference, Oklahoma State University, Stillwater, Okla. (Nov. 1962).

70. Prather, V. B., *The Oil and Gas J.*, **57**, 49 (Nov. 20, 1959).

71. Rohlich, G. A., Robert A. Taft Sanit. Eng. Cent., Rep. No. W 61-3, 130 (1961).

72. Adams, C. E., "Theoretical and Practical Considerations In The Design of Ammonia Stripping Towers." Presented at Environ. Associates Program sponsored by Vanderbilt Univ. (Nov. 1971).

73. Trulsson, S. G., "Ammonia Recovery from Wastewaters in Packed Columns." *J. Water Pollut. Control Fed.*, **51**, 2513 (Oct. 1979).

74. "Fundamentals of Physical-Chemical Processes for the Removal of Nitrogen Compounds from Wastewater." Presented at Nutrient Control Technol. Seminar, Calgary, Alberta, Canada (February 1980).

75. "Optimization of Ammonia Removal by Ion Exchange." Water Pollut. Control Ser., U.S. EPA 17080 DAR 09/71.

76. "Wastewater Ammonia Removal by Ion Exchange." Water Pollut. Control Res. Ser., U.S. EPA 17010 ECZ 02/71.

77. Calvert, C. K., "Treatment with Copper Sulfate, Chlorine, and Ammonia," *J. Am. Water Works Assoc.*, **32**, 1155 (1940).

78. Morris, J. C., "The Acid Ionization Constant of HOCl from 5° to 35°." *J. Phys. Chem.*, **70**, 3798 (December 1966).

79. Ephraim, F., "Inorganic Chemistry," 6th Ed., Interscience Publishers, Inc., New York, N.Y. (1954).

80. Bates, R. G., and Pinching, G. D., "Dissociation Constant of Aqueous Ammonia at 0° to 50° from E.m.f. Studies on the Ammonium Salt of a Weak Acid." *J. Am. Chem. Soc.*, **72**, 1393 (1950).

81. Weil, I., and Morris, J. C., "Kinetic Studies on Chloramine I. The Rates of Formation of Monochloramine, N- Chlormethylamine and N-chlordimethylamine." *J. Am. Chem. Soc.*, **71**, 1664 (May 1949).

82. Saunier, B., and Selleck, R. E., "Kinetics of Breakpoint Chlorination and Disinfection." Report No. 76-2, Sanit. Eng. Res. Lab., Univ. of Calif., Berkeley (May 1976).

83. Morris, J. C., and Wei, I. W., "Chlorine Ammonia Breakpoint Reactions: Model Mechanisms and Computer Simulation". Meeting of Am. Chem. Soc., Div. of Water, Air, and Waste Chem., Minneapolis, Minn. (April 1969).

84. Wei, I. W., and Morris, J. C., "Dynamics of Breakpoint Chlorination," In "Chemistry of Water Supply Treatment and Distribution." A. J. Rubin, (Ed.), Ann Arbor Science, Woburn, Mass. (1974).

85. Saunier, B., and Selleck, R. E., "The Kinetics of Breakpoint Chlorination in Continuous Flow Systems," *J. Am. Water Works Assoc.*, **71**, 164 (March 1979).

86. Stone, R. W., "Full-Scale Demonstration of Nitrogen Removal by Breakpoint Chlorination." MERL, U.S. Environ. Prot. Agency, EPA-600/2-78-029, 69 (March 1978).

87. Stone, R. W., "The Formation of Volatile Halogenated Organic Compounds in Wastewater Chlorination."

88. Selleck, R. E., *et al.*, "Kinetics of Bacterial Deactivation with Chlorine", *J. of the Environ. Eng. Div.*, Am. Soc. of Civil Eng., **104**, EE6, 1197 (December 1978).

89. White, G. C., "Disinfection of Wastewater and Water for Reuse." Van Nostrand Reinhold, New York, N.Y. (1978).

90. Murphy, K. L., "Cost of Nutrient Removal." Presented at Nutrient Control Technol. Sem., Environ. Canada, EPS, Calgary, Alberta, (February 1980).

91. EPA (Burns and Roe Consulting Engineers). "Innovative and Alternative Technology Assessment Manual." U.S. EPA Report No. 430/9-78-009. (1978).

92. CWC Consulting Engineers, "Estimating the Costs of Wastewater Treatment Facilities." Virginia State Water Control Board Rep. (March 1974).

93. Bechtel Inc., "A Guide to the Selection of Cost Effective Wastewater Treatment Systems." U.S. EPA Report No. 430/9-75-002. (July 1975).

94. CWC Consulting Engineers, "Energy Conservation in Municipal Wastewater Treatment." U.S. EPA Report No. 430/9-77-011. (March 1977).

Phosphorus Removal

Removal of phosphorus from waste-waters has been accomplished in municipal plants by chemical, biological, and physical methods. Chemical precipitation using either lime, alum, or ferric chloride is the most widespread method to remove phosphorus and will be discussed in this chapter.

Biological removal is based on the uptake of phosphorus beyond its normal microbial growth requirements by modifications of the activated sludge process. The three modifications to the activated sludge process covered in the following sections are the Bardenpho Process®, the A/O Process®, and the PhoStrip Process®.

Physical removal processes are used to intercept phosphorus from aqueous streams that are to be treated. Three types of processes are discussed here—ultrafiltration and reverse osmosis (both pressure driven membrane processes), and ion exchange.

In examining the various alternatives for phosphorus removal, it is the design engineer's responsibility to carefully consider all aspects of the various design approaches, including:

- Capital costs,
- Chemical costs (local),
- Reliability of process,
- Added labor requirements,
- Impact of the method selected on other wet end operations, and on the solids handling and disposal end, and
- Ultimate disposal of the intercepted phosphorus to prevent accidental re-entry of the phosphate into the waterway from which it was diverted.

CHEMICAL REMOVAL PROCESSES

This section deals with the most commonly used method for removal of phosphorus from waste streams—the addition of certain chemicals that produce insoluble or low solubility salts when combined with phosphate. There are a number of materials that produce comparatively insoluble phosphates. However, the most commonly employed cations are those that minimize cost and produce minimum toxicity in receiving waters or biological systems.

Calcium, aluminum, and iron compounds combine minimum toxicity and minimum cost with maximum effectiveness, which undoubtedly accounts for their widespread use. The compatibility of these chemicals with normal wastewater treatment processes and the familiarity that most design engineers and operators have with these substances also contribute to their popularity. Tables 7.1 through 7.4 show the distribution of municipal treatment plants in the lower Great Lakes basins that remove phosphorus. Of the 154 plants responding to the questionnaire, 104 were using chemicals to remove phosphorus at the time of the survey.[1]

Although the direct purpose of the addition of the three cationic substances discussed is precipitation with phosphorus as a calcium, aluminum, or iron phosphate, there is an abundance of possible side reactions that can involve these three materials. These side reactions can lead to either a more difficult or different sludge handling problem or simply the production of additional sludge. In either event, it must be emphasized that design engineers and operators should thoroughly understand that any operational changes or modifications to the wet end process almost invariably have some impacts on the solids end.

Removal of phosphorus using iron or aluminum salts. This section deals with the most commonly used procedure for phosphorus removal—the use of the readily available trivalent metals, aluminum and iron. Included here is a brief summary of the relevant mechanistic information, along with the basic elements required to design and evaluate the costs of chemical treatment systems and an appraisal of solids handling aspects.

Because there is, as yet, no generally

TABLE 7.1. Number of plants in Lower Great Lakes Basin survey.

Lake Ontario			
New York	48	U.S.	48
Ontario	43	Canada	43
	91		
Lake Erie			
New York	10	U.S.	106
Pennsylvania	1	Canada	32
Indiana	3		
Michigan	23		
Ohio	69		
Ontario	32		
	138		
		Totals: U.S.	154
		Canada	75
			229

accepted view of the mechanisms of phosphate removal using iron or aluminum, the essence of what is known is summarized as it relates to actual practice, which is precipitation and removal of phosphate by chemical means.

The reactions between iron and aluminum salts and phosphate. It is now generally recognized that a metal phosphate is formed when ferrous or ferric aluminum or iron is added to a solution containing phosphate. There are, however, numerous reports in the literature that require a much broader view of the essential mechanisms as they actually occur in practice. These relevant considerations are as follows:

• Optimum pH (5.3 for Fe^{+3} or 6.3 for Al^{+3})[2] generally cannot be used in actual practice although the aluminum opti-

TABLE 7.2. Treatment approaches by chemical (precipitation cation) and location—Lower Great Lakes Basin survey.

	Number of plants								
	U.S.			Canada			Total		
	Al	Fe	Total	Al	Fe	Total	Al	Fe	Total
Primary	1	16	17	2	20	22	3	36	39
Secondary	26	6	32	17	8	25	43	14	57
Tertiary	2	2	4	1	0	1	3	2	5
Total	29	24	53	20	28	48	49	52	101

mum is more attainable than the Fe^{+3} value;

• The equilibrium relationships are shown in the following (Equations 1 and 2):

$$FePO_4 \cdot 2H_2O \rightleftharpoons Fe^{+3} + PO_4^{-3} + 2H_2O \quad (1)$$

where

$$K = 10^{-23}$$

$$AlPO_4 \cdot H_2O \rightleftharpoons Al^{+3} + PO_4^{-3} + 2H_2O \quad (2)$$

where

$$K = 10^{-21}$$

Although these values indicate that a low residual is possible, in practice the mole ratios for a substantial level of removal (greater than 95%) are generally near 1.5 and 2.0. This is probably the result of the necessity to operate at pH values somewhat removed from optimum.

• Phosphate removed with aluminum or iron often settles poorly, requiring the use of excess metal salts or organic polymers, or both, for attaining good settling and, hence, good removal of the phosphate precipitate.[3]

In actual practice, pH is generally higher than the optima shown earlier. According to Leckie and Stumm,[3] this results in the formation of an amorphous compound of Fe^{+3} or Al^{+3} hydroxy phosphate. This postulate has been confirmed[4] insofar as co-precipitation is concerned, although there are mechanistic differences between the postulate of Leckie and Stumm and the findings of others.

Ferrous iron is often used, particularly as waste pickle liquor from the steel industry. Leckie and Stumm[3] indicate that the so called homogenously generated Fe^{+3} is more effective than Fe^{+3} added directly. Hence, Fe^{+2} must first undergo oxidation to Fe^{+3} and then cause precipitation of phosphate in the Fe^{+3} form.

The rate of metal phosphate formation is reported to be virtually instantaneous.[3] In the presence of colloids, metal phosphate reaction is completed first and only afterward does additional metal work to destabilize the colloids and produce hydroxy phosphate compounds. Also, it has been shown that high solids, as in the influent to the primary tank, tend to increase chemical requirements.[5] Increased chemical requirements may be justified by enhanced secondary treatment because of increased removal of colloids in the primary. In addition, it is clear that non-recycle systems will require more chemical than systems using substantial chemical recycle.

In summary, the primary mechanism of phosphate removal by iron or aluminum salts is formation of the corresponding phosphate. As the solubility of the phosphates of iron and aluminum decrease rapidly above 5.2 for iron and above 6.6 for aluminum, excess metal is required. This will amount to 50 to 150% in excess of the stochiometric requirements, although higher required excesses have been reported.[6] Regardless of these considerations, phosphorus removals in excess of 90% and residual phosphorus

TABLE 7.3. Frequency of precipitation cation usage—Lower Great Lakes Basin survey.

Metal	Percent		
	U.S.	Canada	Total
Al	54	41	48
Fe	44	57	50
Lime	2	2	2

TABLE 7.4. Summary of location of phosphorus removal treatment in the treatment plant process—Lower Great Lakes Basin survey.

Process location	Percent		
	U.S.	Canada	Total
Primary	32	46	39
Secondary	60	52	56
Tertiary	8	2	5

(P) values of less than 0.5 mg/L are easily attainable with iron or aluminum with these doses.

Process flowsheets using iron or aluminum salts. Iron or aluminum salts have been used for pre-, post-, and split-treatment techniques, as well as for co-treatment with the activated sludge portion of the treatment facility. These various approaches are shown in Figures 7.1 and 7.2. There are a number of important features associated with each approach that are detailed in the following paragraphs.

Addition to the primary portion of the plant may require a somewhat greater excess of the metal than is usually seen when feeding into the secondary, and such has been reported.[5] Such findings are more likely a result of lack of recycle and inefficient use of metal than of the fact that it is added to the primary.

However, feeding alum or iron to the primary usually results in significantly improved suspended solids and BOD removals, caused by flocculation and adsorption of colloids as well as molecules in excess of 1000 molecular weight. This factor enhances the secondary operation in two ways. First, it tends to underload the aeration system. Second, the enhanced removal of colloids brings about more effective biological treatment.[7] In addition, biological usage can serve to augment removal in primary treatment in a more predictable and definitive way than when phosphorus removal takes place in the activated sludge system. Phosphate can be absorbed by most microorganisms against very large concentration gradients.[8] Hence, microbial function is not altered by reducing the concentration to low (but still adequate) P values in the primary clarifier effluent.

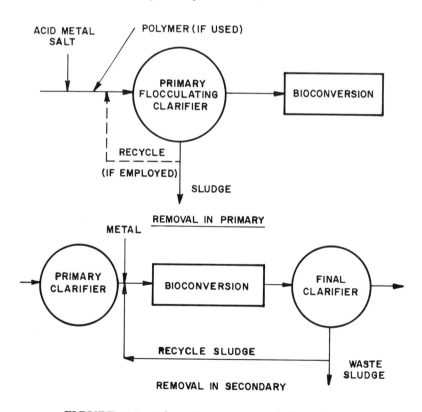

FIGURE 7.1. Phosphorus removal flowsheets.

The principle disadvantage of phosphate removal in the primary clarifier will occur in those instances where a substantial fraction of the phosphate may exist as condensed phosphates. The extent to which hydrolysis is complete at the plant is generally a function of time and wastewater temperature. The size of the collection system affects the hydrolysis level because of its impact on detention time. The fraction of condensed phosphate can be checked easily by the proper analyses. This is an important consideration and one that cannot only increase the required dosage, but also may tend to produce unsatisfactory results with respect to the level of P removal. Condensed phosphate will be higher in the winter.

Finally, co-precipitation or co-sedimentation with primary solids appears to provide good metal phosphate removal. Co-precipitation results in 50 to 60% BOD$_5$ removal and 70 to 85% suspended solids removal. While not practiced, as evidenced by lack of literature reports, recycle around the primary tank will enhance phosphate removal and removal of organics. The applicable theory is contained in the development of von Smoluchowski.[9] Substantial

quantities of data exist on many chemical systems that can validate the concept.[10] Care should be taken to avoid septicity by holding the solids too long, but, short of this, recycle will help.

Removal in the primary tanks has an additional advantage. The chemical conditioning system is therefore at or near the head of the plant where emergency treatment can be best achieved should there be a spill or other heavy dangerous discharge in the sewer system. Also, peak TSS and BOD$_5$ loads are damped by chemical addition to the primaries.

In many activated sludge plants, iron or aluminum is added at the inlet to the secondary portion of the plant. Co-sedimentation with the activated sludge of the metal phosphate appears to be very effective and provides an easy and dependable type of operation.

There are some obvious advantages to the approach using the activated sludge portion of the system. The recycle of a substantial amount of the metal undoubtedly provides a reservoir capable of making up for feed deficiencies. A good portion of the recycled metal is in the hydroxide or hydrated oxide form. This operational buffer undoubtedly ac-

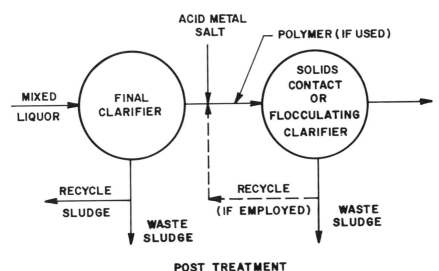

FIGURE 7.2. Phosphorus removal flowsheets.

counts for at least some of the apparent economy of this method. The large inventory of metal in the aeration system will provide great operational security.

The third operating mode is probably the least practiced. This mode involves post treatment in a separate reactor with iron or alum, and an appropriate polyelectrolyte, generally an anionic substance, to assist in the flocculation and sedimentation of the hydroxy phosphate metal combination. This procedure provides excellent removal and a good clean up of the effluent with respect to residual BOD and suspended solids. However, it is somewhat more expensive to practice and requires special wet end capital expenditures that the other procedures do not require. Dow[11] has, or has had, patents on such a process.

Advanced treatment is also feasible for low alkalinity wastewaters that must be nitrified. Addition of alum or iron acid salts to the primary or aeration basin will depress pH and may require lime addition. Also, the lower pH after nitrification will be better for phosphate precipitation. Sand filtration may be used in place of clarification.

Split treatment is also often practiced in conjunction with post treatment. In this case, post treatment becomes a polishing step. The first portion of the split system is usually the primary tank. This treatment may be with either the trivalent metal or with calcium.

For the design engineer, there are several important aspects that must be considered as a part of the design process where iron or aluminum salts are to be used in phosphate removal. These practical considerations are as follows:

• The selection of the best point or points of chemical addition,
• The provision for adequate mixing of added chemicals,
• The provision for the proper mixing and residence time between points of chemical addition, such as for alum and polymers,

• Provision for proper pH control, and
• Provision for the increased sludge volume and change in sludge treatability.

These items will be discussed in detail in the following paragraphs.

Points of Chemical Addition—The points of chemical addition are shown in Figures 7.1 and 7.2. However, the exact point of addition should be selected to assure complete mixing with the wastewater flow. Usually a 90-degree bend or a tee where the chemical addition can be made opposing the flow will assure good, or at least adequate, mixing for alum or iron solutions. This is definitely not the case for polymers and particularly for viscous materials such as anionic or nonionic organic polyelectrolyte. If the polymer cannot be added to the mixing zone of a highly mixed or internally recirculated clarifier, it should be introduced preceeding a dynamic or static mixer. Usually only a few seconds (10 to 30) are needed to mix most polymers. Both nonionics and anionics will suffer badly from excessive mixing, and process efficiency will fall off. Cationics are much more "forgiving" with respect to excess mixing. Poor mixing results in excess polymer requirement, poorer settling, and poor thickening.

Mixing efficiency can be increased and the required mixing power reduced by reducing the polymer concentration with dilution water at or near the point of addition. The polymer should be added at a concentration of 0.01 to 0.05%, and never above 0.1%. Secondary effluent is satisfactory for final dilution at point of usage, but not for making up the polymer solution. However, care should be taken to manage the concentrations in such a way as to maximize the shelf life of the polymer. Be guided by the polymer manufacturer's recommendations.

Considerable research on water and wastes[10] where inorganics and organics or two organics are required for good flocculation has shown that 10 seconds between addition and mixing of the first chemical and the second chemical is a practical minimum. Presumably, this represents the time after 10 to 30 seconds of mixing to permit maximum contact between chemical and solids. Shorter times will result in some increases in chemical requirements.

The use of iron or aluminum at near neutral pH values produces, through hydrolysis, a strong acid which, in turn, reacts with some of the available alkalinity. Streams with low pH or low alkalinity may require pH trimming with lime or some other source of alkalinity, such as sodium aluminate or a combination of alum or aluminate.

Lime feeding (discussed in a later section) represents an effective, low-cost solution. Operationally effective lime feed systems require careful design, as does any system that involves the feeding of a slurry.

Regardless of how it is controlled, some pH control is often necessary. This problem is magnified in those cases where nitrification is also practiced. Careful appraisal of background alkalinity, dosage requirements, and downstream impact is required. However, if it is necessary to increase the pH at the aeration basin because of low alkalinity, then postaddition of metal salts between aeration and clarification should be considered.

Sludge volumes and quantities will be increased by the addition of metals for phosphate removal. Typically, with a 200 mg/L BOD and suspended solids raw wastewater, there will be an increase in sludge volume of 30 to 50% with 100 mg/L of alum or of its iron equivalent. This will result in a 10 to 20% increase in dry solids mass of the total sludge produced by wastewater treatment. The lower volume increases occur when the addition is made to the primary tank as opposed to the greater volumetric increases when the metal is added to the aeration basin.[12-14] These values can range widely, particularly the volume changes. Other important effects occur downstream in the sludge handling system that must be considered when the sludge conditioning system is being designed. The presence of iron or aluminum in the sludge invariably reverses the charge requirement for conditioning prior to dewatering. However, the situation is complex and the chemical requirement changes as the ratio of wastewater sludge to metal precipitates changes. Polymer use on the wet end also compounds the polymer requirements for sludge dewatering. The best solution to this complex situation is to plan for dual polymer feed systems, each of which can be used to feed cationic, anionic, or nonionic polymer at maximum rates of 12.5 g/kg (25 lb/ton) conditioned dry solids. This is a critical problem that, if ignored in the original design, usually requires a difficult and expensive retrofit.

Performance summary. Tables 7.5 and 7.6 contain representative data for systems using iron or aluminum in one or more of three operating modes described in the text.

Phosphorus removal with lime. This section of the chapter deals with one of the earliest methods of chemical precipitation used for phosphorus removal, that is, the precipitation of relatively insoluble compounds of calcium and phosphate. Preference was given to lime precipitation by early researchers in the field[22-24] because of low cost, ease of handling, low corrosion potential, and operator familiarity. Additionally, lime tended to work well under certain treatment conditions.

Rudolph[22] was one of the first to study and quantify the essential aspects of the process. Since that time, and particularly in the 1960s, there has been frequent reference to the use of lime for phosphate removal.[25-30]

Calcium reactions with phosphate. There are five calcium phosphate compounds that can be precipitated from

143

TABLE 7.5. Performance information on systems using Fe^{+3} or Al^{+3}.

Point of metal addition	Fe^{+3} or Al^{+3} quantity, mg/L	Effluent P or removal, mg/L	Plant size	Reference or comment
Primary	Fe = 14.4	1.7 (83%)	9 mgd	Michigan City, Ind.[15]
	Al = 7.2	1.0 (89%)	9 mgd	Michigan City, Ind.
Aeration basin	Fe = 13.9	0.82 (90%)	9 mgd	Michigan City, Ind.
	Al = 5.5	0.80 (93%)	9 mgd	Michigan City, Ind.
Primary	Al = 7.0	1.00 (87%)	24 mgd	0.4 mg/L Polymer Windsor, Ontario[16]
Aeration basin	Al = 5.0	0.70 (83%)	15 mgd	Monroe County, N.Y.[17]
End of aeration basin	Al = 4.2	0.90 (82%)	8.6 mgd	Columbus, Ind.[18] Anionic polymer
After aeration	Al = 3.0	0.80 (88%)	15 mgd	Gates, Chili Ogden[19]
Primary	Al = 3.7	1.50 (61%)	90 mgd	Frank Van Lare, Polymer[19] Rochester, N.Y. 0.4 mg/L
After final post treatment	Fe = 8.7	0.90 (86%)	3 mgd	Big Sister Creek Polymer[19] Angola, N.Y. 0.6 mg/L
After trickling filter	Al = 2.5	0.60 (85%)	0.9 mgd	Ely, Mich.[19]
Post treatment	Al = 18.0	0.60 (90%)	Bench	Canton, Ohio[20]
	Fe = 24.0	2.10 (70%)	Bench	Canton, Ohio

aqueous solutions, as reported by Van Wazer.[31] These are:

1. $Ca(H_2PO_4)_2$,
2. $Ca(H_2PO_4)_2 \cdot H_2O$,
3. $CaHPO_4$,
4. $CaHPO_4 \cdot 2H_2O$, and
5. $Ca_5(OH)(PO_5)_3$ (highly variable extent of hydration).

The solubility product varies over a range of 10.[11] Clark[32] reports the $K_{25°}$ as follows:

$$K_{25°} = [Ca]^{10} [PO_4]^6 [OH]^2 = 10^{-115} \quad (3)$$

Although, as indicated in an earlier section, other phosphate forms (termed condensed) are found in raw wastewater, this discussion is limited to orthophosphate—the principal form hydrolyzed in sewers and treatment processes.

Menar and Jenkins[25] also confirm that the significant form is hydroxyapatite $[Ca_5(OH) (PO_5)_3]$ "at all encountered sewage process pH values." They report that pH, magnesium concentration, carbonate (or alkalinity), and fluoride all significantly affect hydroxyapatite precipitation. High pH values increase the rate of precipitation, while magnesium and carbonate appear to inhibit hydroxyapatite formation. It is unlikely that precipitation, as it is normally practiced, will be affected by these factors to the extent of requiring a change in basic process design, but significant second-order effects have been reported.

Related processes using lime have been described by Bogan et al.,[33] Menar and Jenkins,[25] Jebens and Boyle[34] and others.[35,36] In these instances, the removal of phosphate that occurs as a result of a pH increase associated with photosynthesis or a high rate of aeration in either suspended growth of fixed film systems is noted. These precipitations occur because of the interaction between background calcium and waterborne phosphate.

Rudolphs,[22] in the second of four papers on the subject, related alkalinity, pH, turbidity, and lime addition in a study of wastewater clarifications with lime. Sawyer[23] found that a pH of 11 was required to reduce phosphate to residual values of less than 0.5 mg/L phosphorus. Bogan et al.[33] observed that

TABLE 7.6. Summary of some results from full-scale treatability studies (Ontario Studies).

| Chemical[a] | Total no. of plants | Addition to raw wastewater | | | | Addition to mixed liquor | | | |
| | | Number of plants | Avg. dosage (mg/L) | Avg. molar ratio metal ion to: | | Number of plants | Avg. dosage (mg/L) | Avg. molar ratio metal ion to: | |
				TP[e]	FP[f]			TP[e]	FP[f]
Ferric chloride	27	7	14.2[b]	2.7	4.2	20	9.5[b]	1.5	2.3
Alum	20	5	10.3[b]	1.7	—[d]	15	7.5[b]	1.6	2.1
Lime	4	4	10.1[c]	15.8	—[d]	—	—	—	—

[a] Effluent total phosphorus objective of 1 mg/L
[b] Calculated as Fe or Al
[c] Calculated as Ca
[d] Insufficient data
[e] TP = Total Phosphorus
[f] FP = Filterable Phosphorus

at high pH (greater than 10), the free calcium ion concentration substantially affected residual phosphate. Schmid & McKinney[29] obtained data relating phosphate removal to lime dose and pH in the treatment of raw wastewater.

Quantities of lime required. The quantity of lime required for phosphorus precipitation is proportional to the buffering capacity of the system and is not related to the quantity of phosphate present in most wastewaters. Most researchers[27,30,37] report that removal of phosphate to values of less than 1.0 mg/L phosphorus requires pH values of 10.5 to 11.0. Lime requirement is also related in part to magnesium concentration, although the removal of calcium as the carbonate is a more important competitive reaction. Wuhrman[26] has stated that the lime requirement for reaching a pH of 10.5 to 11.0 will be 1.5 times the carbonate hardness. Buzzell and Sawyer[30] obtained data on several wastewaters on the lime requirement to reach 11.0. These data tend to support Wuhrman's generalization.

Pertinent findings have been obtained at the South Tahoe PUD Plant[38] where 400 mg/L of lime was required to raise the pH to 11.5. This, in turn, resulted in a residual phosphate of 0.3 mg/L after sedimentation. Similar findings have been reported at Blue Plains, Washington, D.C.;[39] Pomona, Calif.;[40] and Lebanon, Ohio.[41]

There is some conflict concerning the settleability of the precipitates formed at high pH.[37,39] Authors have suggested using coagulants, flocculants, or filtration following the precipitation and sedimentation step. Extensive studies[10] of calcium precipitate stability indicate that these apparently conflicting observations probably result from substantial differences in mixing regimes and the extent of internal and external recycle from application to application. Extensive recycle, regardless of mode, will result in a stable system capable of rapid and complete sedimentation.

Granular media filtration of slurries or clarifier overflows at high pH values and with a substantial calcium excess should be undertaken only with caution. Such caution is required because of the cement-like character of the combination of the precipitates and the filter media. This precipitation is facilitated by the interparticle contact brought about by the nature of the flow in a granular media filter.

Process flowsheets employing lime. Lime traditionally has been used either as a precipitant in advanced treatment operations or in the primary portion of the treatment works where co-precipitation with organics assists in removal. The P/Ca ratio is the highest in the primary and, importantly, the biological uptake associated with synthesis is not concentration-dependent in the PO_4^{-3} range of concentrations under study. Therefore, P can be removed under the most advantageous circumstances.

Schmid and McKinney[29] proposed the use of the primary system in their bench-scale study of phosphate removal. Their proposed flowsheet is shown in Figure 7.3. The system uses flocculation, but not recycle of solids. Hence, flocculation must be very efficient to ensure a stable system. The presence of organic solids undoubtedly assists in stabilization and precipitation. Significantly, lower pH values and, hence, lower lime doses are required for the same removals when the biological system is used as the second stage of phosphate removal, rather than as the first. This results from the lack of concentration dependence shown by biological systems.

There is a substantial concentration dependence shown where lime is used for phosphorus precipitation.[23,29,33] In general, the lime requirement to remove phosphate from 2.5 mg/L to a concentration of 1.5 mg/L is three times that required to reduce the phosphate concentration from 5 to 4 mg/L.

Dorr Oliver, in its PEP™ process, appeared to take advantage of dosage considerations. External recycle, which markedly increased the efficiency of lime, was used (Figure 7.4). The recommended operating pH values for the PEP system vary between 9 and 10.[42] The lime dosage was about one half that required for advanced treatment for the same results. A solids concentration of 500 to 2000 mg/L was maintained in the recycle system.

A further advantage of the application of lime in the primary system is the underloading of the secondary through increased BOD removal. The BOD removal in a primary unit process operated at a pH of 9.5 will be on the order of 50 to 60%. Care must be taken to match the quantity of organic material remaining after the primary with the desired effluent concentrations of phosphate.

Lime treatment in the primary portion of the plant results in an increase in primary sludge of about 100%.[29] This increase will rise no further if the system

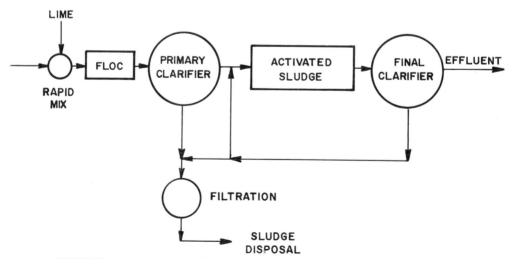

FIGURE 7.3. Process flowsheet for lime removal in the primary.

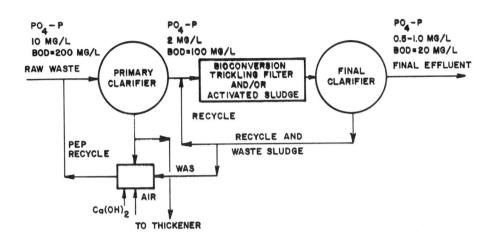

FIGURE 7.4.   The PEP® system for phosphorus removal.

is operated at pH values that do not markedly affect alkalinity, that is, at levels less than 9.5. If higher pH values are used, the quantity of primary sludge will be a function also of the calcium carbonate precipitated.

The alternate procedure for the use of lime is an advanced treatment step. When practiced in this fashion, the process is identical to a softening operation with magnesium precipitation. The operating pH, as described earlier, is on the order of 11.0 and, as a consequence, removal of magnesium and carbonate is essentially complete, with substantial removals of silica and fluoride also occurring. A typical flowsheet for advanced treatment with lime is shown in Figure 7.5.

The sludge from the process will behave much like sludge from a softening process. If the sludge is principally $CaCO_3$ (greater than 90%), it will thicken at 195 to 293 $kg/m^2 \cdot d$ (40 to 60 lb/d/sq ft)

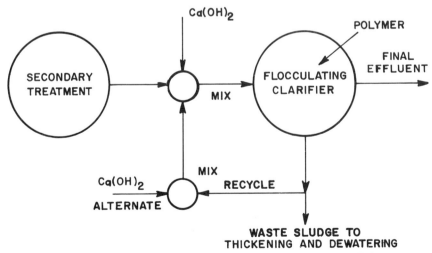

FIGURE 7.5.   Advanced wastewater treatment with lime.

to approximately 20 to 30% solids, and will filter at 98 to 244 kg/m$^2 \cdot$h (20 to 50 lb/hr/sq ft) to 40 to 60% solids. With increasing magnesium, the thickening and filtration rates will decrease. Typical values for sludge containing 20 to 40% Mg(OH)$_2$ by weight will be for thickening 49 to 98 kg/m$^2 \cdot$d (10 to 20 lb/d/sq ft) to 15 to 30% solids and filtration rates of 49 to 98 kg/m$^2 \cdot$h (10 to 20 lb/hr/sq ft) to a cake solids of 25 to 40%. The hydroxyapatite will behave much like aluminum or ferric hydroxide in terms of thickening or dewatering.

Where the alkalinity is high (greater than 200 mg/L as CaCO$_3$), special consideration should be given to the thickening behavior of the precipitation reactor. The sludge can become unmanageable and may require such a low inventory that scaling and low phosphate removal results. This situation can be remedied, all or in part, by the use of starch-based polymers for "fluffing" the precipitate or for external recycle that artificially increases floor loading and provides seed crystals, as suggested by Leckie and Stumm[2] and Ferguson.[36]

Others have noted this aspect of calcium salt precipitation with CaSO$_4$ and CaF$_2$.[10] Depending on the precise circumstances, one or more of the alternates shown in Figure 7.6 can be used.

The quantity of sludge to be recycled must be determined or estimated from the thickening behavior of the system. However, it generally will fall in the range of 5 to 20% of the inflow. This applies to those cases where the underflow is pumped at 10 to 20% solids. There is, however, no substitute for knowing the real loading and underflow concentration relationships and their relationships to the pumpability of the slurry. These data can be obtained from bench-scale studies.

In most cases, recycled flow (Figure 7.6) is discharged, either after lime addition or with the lime, into the influent line. If polymers are to be used with external recycle, a practical minimum of 10 seconds between lime and polymer addition to maximize polymer effectiveness has been suggested.[10] Generally speaking, the polymers effective at high pH and in the presence of excess lime

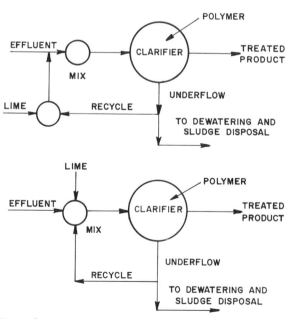

FIGURE 7.6. Recycle procedures to stabilize calcium precipitation systems.

TABLE 7.7. Performance information of systems using lime.

Primary or post treatment	pH	Ca(OH)$_2$ dose (mg/L)	Effluent phosphorus (mg/L)	Scale or size	Reference
Post	10.5	400	0.1 to 0.6	Full	Davis (South Tahoe)[38]
Post	11.0	545	5.2 (unfiltered) 0.4 (filtered)	Full	Owen[24]
Post	—	300	2.3	Pilot	Eberhardt & Nesbit[43]
Primary	9.5	150	2.0	Bench	Schmid et al.[29]
Either	8.0	—	85 to 90% removal	Bench solids recycle	Ferguson et al.[36]

will be anionic in character, but only bench studies will serve to optimize polymer type and dose.

Performance of lime precipitation system. Table 7.7 contains a summary of relevant performance information on systems using lime for precipitation of phosphate from wastewater. This table includes data obtained on bench-, pilot- and full-scale tests for both types of flowsheets discussed earlier.

Design data for chemical systems. This section contains clarifier, thickener, and filter loading data for the chemical treatment systems used in phosphate removals. The values given are meant to provide guidance and have been found in practice to produce satisfactory results. Most of these values have a substantial safety factor in-

cluded to handle such imperfections as a low hydraulic efficiency (35 to 60%) in a clarifier. Regardless of this, the design engineer should, at every opportunity, use bench or pilot studies to confirm or augment design data obtained from other installations.

Table 7.8 contains design information for clarification.[44] Minimum side-water depth for circular clarifiers, when treating with Fe^{+3} of Al^{+3}, should be 3.66 m (12 ft). A side-water depth equal to 1.6 times the square root of the clarifier diameter ($SWD = 1.6 \; D^{1/2}$) will provide a good SWD:D design. Feed well diameters should be on the order of 30 to 40% of the tank diameter, and depths of 60 to 75% of the tank depth. Minimum bottom slopes towards the point of sludge withdrawal should be 1:12. Tanks should be equipped with scum removal equipment.

TABLE 7.8. Clarifier loading rates.[a]

Type of treatment	Design loading, L/m$^2 \cdot$s (gpm/sq ft) conventional clarifier polymer		Design loading, L/m$^2 \cdot$s (gpm/sq ft) blanket clarifier polymer	
	No	Yes	No	Yes
Lime in primary clarifier to pH 9–10.0	0.4 (0.6)	0.8	0.6	1.0
Fe^{+3} or Al^{+3} in primary clarifier	0.3 (0.4)	0.6	0.5	0.8
Fe^{+3} or Al^{+3} aeration basin final clarifier	0.3 (0.5)	N/A	N/A	N/A
Post treatment with lime to pH greater than 10.0	0.4 (0.6)	0.8	1.0	1.5
With Al^{+3} and Fe^{+3}	0.3 (0.4)	0.6	0.6	0.8

[a] Assumes a minimum SWD of 3.6 m (12 ft), and SWD = 1.6 D$^{1/2}$ ("English" units).

In general, the same recommendations apply to comparable items on rectangular tanks, particularly with respect to depth, inlet velocity control, and bottom slope. Conventional weir loading criteria should be used on both types of tanks.

Solids thickening data are given in Table 7.9. Loading rates and results assume that the displacement time in the thickener for biological and primary sludges is managed to prevent gasification and subsequent decrease in solids concentration because of flotation/mixing.

The polymer requirements for sludge conditioning associated with dewatering are complex and should be determined on bench scale. There are some basic considerations that can be used as a guide in preliminary design and in performing bench-scale work. These are:

• When studying requirements on bench scale, the actual operating conditions should be simulated, particularly with respect to sludge age, septic state, and fraction of WAS and primary sludge;

• In general, polymer requirements will fall between 2.0 to 2.5 g/kg (4 and 5 lb/ton) on the low side, and 7.5 g/kg (15 lb/ton) on the high side in conditioning a sludge of biological origin or a mixed sludge;

• When chemicals are used on the wet end, whether trivalent metals or organics, it is possible that two polymers will be required for sludge management. Daily, and sometimes twice daily, bench tests will be required to optimize polymer doses.

BIOLOGICAL PHOSPHORUS REMOVAL

Phosphorus in raw wastewaters exists in three forms: orthophosphorus (PO_4), polyphosphates (P_2O_7), and organic phosphorus. Orthophosphorus can be assimilated readily by microorganisms, and the polyphosphates and organic phosphates are usually hydrolyzed by microorganisms to the ortho form. The polyphosphate and organic phosphorus content may account for 40 to 70% of the phosphorus in wastewaters. Secondary biological treatment systems accomplish some phosphorus removal by the use of phosphorus for cell synthesis during BOD removal. A typical phosphorus content in sludge from biological systems is in the range of 1.5 to 2%, based on dry weight. The amount of phosphorus removed by sludge wasting may be in the range of 10 to 30% of the influent amount. This is affected by sludge handling techniques and side stream return flows.

Phosphorus is an important element for microorganisms because of its use in energy transfer and for cell components

TABLE 7.9. Thickener design criteria and polymer requirement.

Solids origin	Chemicals used	Gravity thickener loading rate and results, kg/m^2·d (lb/d/sq ft)	Solids (%)	Flotation thickener loading rate and results L/m^2·s (gpm/sq ft)	kg/m^2·h (lb/hr/sq ft)	Solids (%)/ polymer Conditioning
Primary	Fe^{+3}, Al^{+3}	20–29 (4–6)	4–5		N/A	
Primary	Lime	20–39 (4–8)	5–8		N/A	
Activated sludge	Fe^{+3}, Al^{+3}	15–20 (3–4)	1.5–2.0	0.7–1.0 (1.0–1.5)	10–24 (2–5)[c]	2.5–4.5 Cationic 5–10 lb/ton Anionic 2–4 lb/ton
Post precipitation	Fe^{+3}, Al^{+3} Lime[b]	15–24 (3–5) 150–240 (30–50)	2.0 20–40	0.7–1.0 (1.0–1.5)	10–24 (2–5)[c] N/A	3.0–5.0 Anionic 1–4 lb/ton Anionic 1–4 lb/ton

[a] Includes recycle of 75 to 100% of Q.
[b] Depends on Ca/Mg.
[c] Is controlled in part by recycle rate and air/solids ratio. The air/solids ratio should be of the order of 0.02 lb/lb.

such as phosopholipids, nucleotides, and nucleic acids. Attachment of an ortho-phosphate (Pi) radical bond to adenosine diphosphate (ADP) to form adenosine triphosphate (ATP) results in the storage of energy (7 kcal/mole P), which is available on conversion of ATP to ADP.

$$ATP \rightleftharpoons ADP + Pi + energy \qquad (4)$$

Phosphorus is also contained in nucleotides such as nicotinomide adenine dinucleotide (NAD) and flavin adenine dinucleotide (FAD), which are used for hydrogen transfer during substrate oxidation-reduction reactions. The nucleic acids, ribonucleic acid (RNA) and deoxyribonucleic acid (DNA), are composed of either a ribose or a deoxyribose sugar structure with attached bases of adenine, cytosine, guanine, and thymine. The sugars are linked by phosphorus bonds. Phosphorus may account for 10 to 12% of RNA or DNA molecular weight.

In 1955, Greenberg et al.[45] proposed that activated sludge could take up phosphorus at a level beyond its normal microbial growth requirements. Srinath[46] reported on batch experiments in 1959 to conclude that vigorous aeration of activated sludge could cause the concentration of soluble phosphorus in mixed liquor to decrease rapidly to below 1 g/m^3. Levin and Shapiro[47] used the term "Luxury Uptake" of phosphorus to describe high levels of phosphorus removal on vigorous aeration of a municipal wastewater. Over 80% phosphorus removal was achieved in their experiments. They reported that a small amount of 2-4 di-nitrophenol inhibited the phosphorus uptake, indicating the removal was of biological origin. They also observed volutin granules in the bacterial cells, which are reported in the microbiology literature to contain polyphosphates. Upon anaerobic conditions, high levels of phosphorus leakage from the biological cells occurred. Barnard observed the need for anaerobic-aerobic cycling for biological phosphorus removal within the activated sludge system.[48]

In the following sections, three biological phosphorus removal systems will be discussed: the PhoStrip Process, the A/O Process, and the Bardenpho Process.

There are two concepts involved in these three proprietary processes: the design basis for the modified Bardenpho Process and the A/O Process is mainstream biological phosphorus removal via sludge wasting; and the PhoStrip Process makes use of side-stream leaching.

Figure 7.7 shows one of several variations of each of the three patented biological phosphorus removal systems. The PhoStrip system takes advantage of biological phosphorus uptake and release mechanisms to concentrate the released phosphorus in a relatively small stream (5 to 25% of the influent flow). Lime treatment is then applied for phosphorus precipitation and ultimate phosphorus disposal via the lime sludge. In addition, some phosphorus is removed with the waste activated sludge.

The Bardenpho system was developed for nitrogen removal and later modified for phosphorus removal. An anaerobic contacting stage ahead of the first denitrification stage conditions the return sludge for biological phosphorus uptake in the aeration zone.

The A/O system, however, applies an anaerobic-aerobic contacting sequence to develop a biological population that can accomplish high levels of phosphorus removal via the wasted sludge. Because biological nitrification and denitrification for nitrogen removal is not an essential feature of the A/O system, a higher rate activated sludge system can be achieved with the A/O system than with the Bardenpho system. This results in a higher sludge production and thus greater quantities of biological phosphorus removal per unit of BOD removed for the same fraction of phosphorus in the waste sludge. Both the A/O and PhoStrip systems have alternative flow schemes that include nitrogen control (Chapter 8).

The PhoStrip Process. The PhoStrip Process is an activated sludge process

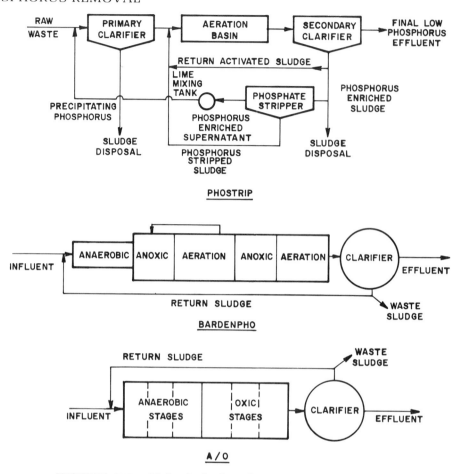

FIGURE 7.7. Biological phosphorus removal systems.

designed to accomplish phosphorus removal. As is shown in Figure 7.5, this process differs from conventional activated sludge in that either all or a portion of the return sludge is subjected to "phosphorus stripping" by exposing the sludge to anaerobic conditions in a stripper tank. The solids retention time (SRT) in this tank typically ranges from 8 to 12 hours. During this anaerobic period, phosphorus is released and is elutriated from the sludge in the stripper tank continuously by a stream that is low in phosphorus content. This stream may be either the overflow from the chemical treatment tank (Figure 7.5) or primary effluent, or the underflow sludge from the stripper tank itself. The phos-

phorus rich overflow from the stripper tank passes continuously to the chemical treatment tank where lime is added for phosphorus precipitation. Because of the flexibility associated with the portion of recycled sludge that can be subjected to anaerobic conditions for different detention times in the stripper tank, a wide range of phosphorus removal levels can be achieved.

Masse[49] reported that the following factors are considered in design of a PhoStrip system: influent characteristics (phosphorus loadings, temperature, pH, alkalinity, hardness, and total dissolved solids); effluent requirements for phosphorus, N, SS, and BOD; mainstream operation mode (hydraulic retention time in aeration, MLVSS, MLSS, design RAS flow, and WAS generation); potential stripper feed NO_2-N and NO_3-N concentration; and sludge processing objectives and recycle liquor characteristics.

PhoStrip systems operate with conventional, tapered, step, modified, and pure-oxygen activated sludge modes without interfering with optimum BOD and suspended solids removal objectives. Either plug or mixed flow reactors can be used. The PhoStrip Process reportedly adapts to any aeration rate, food-to-microorganism ratio, and sludge recycle percentages used in the various activated sludge modes. PhoStrip systems have operated successfully with MLSS concentrations ranging from 600 to 5000 mg/L. Basically, if between 1- and 10-hour retention times are pro-

vided in the aeration basin, the conventional PhoStrip Process can be applied. With modification, the PhoStrip Process is also applicable to contact stabilization and extended aeration modes.

Operational data for several full-scale PhoStrip facilities are shown in Table 7.10.

The A/O Process. The A/O Process is an activated sludge process primarily designed to accomplish biological phosphorus removal. As is shown in Figure 7.5, three anaerobic stages (the "A" in A/O) are followed by three or more aerobic or oxic stages (the "O" in A/O). Recycle sludge from the secondary clarifier is mixed with either raw wastewater or primary effluent in the anaerobic stage so that there is sorption of

TABLE 7.10. Summary of full-scale PhoStrip performance data.

| PhoStrip site demonstration (year) | Flow | | Month | Average phosphorus concentrations (mg/L) | | | | | |
| | | | | Raw influent | | Primary effluent | | Final effluent | |
	L/s	mgd		Total	Ortho	Total	Ortho	Total	Ortho
Seneca Falls, NY[a] (1973)	39	0.9	7–8	6.3				0.6	
Reno/Sparks, NV[b] (1974–1975)	260	6	8/1974 to 6/1975			7.0			1.0
1977[c]	530	12	1	9.1				0.9	0.5
1982[d]	1050	24	9	7.5				1.0	0.5
			9–11	7.0				1.0	0.5
Carpentersville, IL[e] (1978–1979)	220	5	12		5.4		4.0		0.5
			1		5.8		5.2		0.5
			2		3.4		3.4		0.5
			3		3.3		7.1		0.5
Adrian, MI[e] (1981)	310	7	6	4.3		3.6		0.5	
			7	4.5		4.9		0.6	
			8	5.5		6.3		0.3	
			9	4.1		3.6		0.3	
			10	4.2		4.4		0.5	
Southtowns, NY[e] (1982)	700	16	3	2.3					0.4
			4	2.5					0.3
			5	3.8					0.4
			7	4.1					0.9
Savage, MD[e]	660	15	5	8.4	5.4	8.5	5.6	1.0	0.8
			6	9.4	5.5	9.3	5.1	1.0	0.4
Lansdale, PA[e] (1982)	110	2.5	5–8	5.8				0.9	
Amherst, NY[e] (1982)	1050	24	8–10	4.1				0.4	

Sources: [a]Levin *et al.* (1975); [b]Peirano (1977); [c]Drnevich (1979); [d]Peirano (1982) and Peirano *et al.* (1982); [e]data provided by indicated municipality.

BOD by the organisms with accompanying phosphorus release necessary for biological phosphorus removal. Anaerobic conditions in the initial contact zone are critical to the overall phosphorus removal efficiency of the A/O Process (as distinguished from both the PhoStrip and Bardenpho systems). As a result, the soluble BOD_5 concentration in the entering wastewater must be of sufficient strength to ensure the absence of all exogenous electron acceptors (both oxygen from the atmosphere and oxidized nitrogen, if any, from the recycled sludge). Low concentrations of exogenous electron acceptors are assured by either providing a cover or sparging with nitrogen gas. The oxic stage, essential for the metabolism of BOD_5 and uptake of the phosphorus released in the anaerobic stage, may be aerated with either air or oxygen. Phosphorus is removed from the system in the waste sludge, which may contain 4 to 6% phosphorus by dry weight.

A/O Process design considerations. Suggested formats for baffled staging of the A/O system include: three anaerobic stages, three anoxic stages, and four aerobic (oxic) stages. Recommended mixed liquor suspended solids levels are around 2000 mg/L, with design organic loadings between 0.15 and 0.7 kg BOD_5/kg MLSS·d. Recognizing the distinction between soluble particulate and filterable influent organics, a soluble organic loading range of 0.08 to 0.4 kg BOD_5/kg MLSS·d was proposed. Acceptable temperature extremes are 5° to 30°C. The ratio of BOD_5/P may limit the removal of P or control the residual P level.

Sizing the composite reactor stages primarily depends on hydraulic detention times. Recommended values for the anaerobic and anoxic stage are between 0.5 and 1.0 hour for each segment. The oxic stage size depends on nitrification requirements. Without oxidation of nitrogen, values of 1.8 to 2.5 hours are recommended. The oxic stages must be enlarged for nitrification, however, with recommended values of 3.5 to 6.0 hours.

Relatively low mixing power inputs of 9.3 kW/1000 m^3 (0.35 hp/1000 cu ft) of tank volume were suggested for the anaerobic and anoxic segments. In the case of the oxic stage aeration, excess aerator capacity was recommended. Basin and multiple aerator configurations must, however, be amenable to partial shut downs or reduced output to facilitate a match between available oxygen transfer and existing demand.

The final clarifier system should be developed according to a 1.02 to 1.10 m/h (600 to 650 gpd/sq ft) overflow rate and a desired sludge blanket depth below 0.6 m (2 ft). Attention to expediting bottom sweep should help to mitigate phosphorus-bleedback in this area. Operating design return sludge recycle values average 30% (based on influent flow). However, this may be as high as 60% depending on design MLSS concentration and return sludge concentration.

Operating data from three operating phases of pilot A/O Processes are given in Table 7.11. Full-scale performance data for the Largo, Fla., plant are shown in Table 7.12.

The Bardenpho Process. The Bardenpho Process is an activated sludge process designed to accomplish biologically both nitrogen and phosphorus removal. As shown in Figure 7.5, two anoxic stages are used to accomplish high levels of biological nitrogen removal by denitrification. A fermentation stage is added ahead of the four-stage Bardenpho nitrogen removal system to create anaerobic-aerobic contacting conditions necessary for biological phosphorus uptake. Phosphorus is removed from the system in the waste sludge, which contains 4 to 6% phosphorus by dry weight. Depending on the relative amounts of phosphorus, BOD_5, and nitrogen in the influent, very low levels of phosphorus (less than 1 mg/L) can be achieved in the effluent. For weaker wastewaters or high influent phosphorus concentrations, a small amount of chemical, such as alum or ferric salts, is added to polish the effluent phosphorus to below 1

TABLE 7.11. Performance of A/O pilot plants.

Location	Rochester, NY	Rochester, NY	Largo, FL
Process type	A/O	Nitrifying A/O	A/O
Retention time (hours)			
Anaerobic	0.5	1.0	0.5
Aerobic	1.5	3.0	1.6
MLVSS (mg/L)	3850	4470	2650
F/M kg BOD/kg·d MLVSS	0.58	0.24	0.66
(lb BOD/lb MLVSS/d)			
Influent			
BOD (mg/L)	186	184	155
SS (mg/L)	314	219	144
Total PO_4-P (mg/L)	4.9	7.5	9
NH_3-N (mg/L)	—	16.9	—
Effluent			
BOD (mg/L)	22.1	19.6	13
SS (mg/L)	40	26	16
Filtered PO_4-P (mg/L)	0.49	0.38	0.95
NH_3-N (mg/L)	—	0.50	—
Return sludge concentration (%)	3.81	3.56	1.81
SVI (mg/L)	38	52	67

mg/L, if required. Because of the liquid detention times and SRTs required for nitrification and denitrification, a relatively high-quality effluent in terms of BOD, suspended solids, and ammonium nitrogen concentrations is possible. The resultant SRT provides an aerobically stabilized sludge that has been disposed of without further stabilization.

Table 7.13 shows typical Bardenpho stage detention times. These are affected by wastewater temperature,

TABLE 7.12. Performance of Largo A/O plant.

	Flow regime	
	Low flow	High flow
Parameter	Mean ± std. deviation	
Flow rate L/s (mgd)	140 ± 9 (3.2 ± 0.2)	175 ± 4 (4.0 ± 0.1)
Retention time (hours)		
Anaerobic	90 ± 6 (2.05 ± 0.14)	54 ± 1 (1.24 ± 0.03)
Aerobic	90 ± 6 (2.05 ± 0.14)	90 ± 3 (2.06 ± 0.06)
MLVSS (mg/L)	2540 ± 250	2560 ± 300
F/M kg BOD/kg·d MLVSS	0.26 ± 0.07	0.36 ± 0.06
(lb BOD/lb MLVSS/d)		
Influent		
BOD (mg/L)	113 ± 26	127 ± 16
SS (mg/L)	92 ± 29	147 ± 40
Total PO_4-P (mg/L)	8.32 ± 0.97	9.5 ± 0.95
Effluent		
BOD (mg/L)	8.3 ± 4.5	9.8 ± 3.0
SS (mg/L)	12.3 ± 3.0	20.3 ± 7.5
Filtered PO_4-P (mg/L)	0.85 ± 0.53	0.80 ± .42
Clarifier overflow m^3/m^2·d (gal/d/sq ft)	25.8 (635)	32.6 (800)
Return sludge concentration (%SS)	1.51 ± 0.25	1.71 ± 0.26
Waste activated sludge kg/d (lb/d)	890 (1970)	1950 (4300)

TABLE 7.13. Typical Bardenpho process detention times.

Stage	Detention time (hours)
Anaerobic	1–2
First anoxic	2–4
Nitrification	3–8
Second anoxic	2–4
Reaeration	0.5–1

BOD$_5$ concentration, total nitrogen concentration, effluent requirements, and sludge handling considerations.

The Bardenpho Process design approach must evaluate design requirements for each of the five stages to accomplish phosphorus removal, nitrification, and denitrification. Sludge disposal considerations may affect the process design sizing. For example, many Bardenpho system designs based strictly on achieving nitrification and denitrification may result in a final SRT that may be within the range of sludge stabilization by aerobic digestion. In such cases, design SRT values are increased by increasing the detention times of the aeration tanks to achieve a stable sludge (thus eliminating separate sludge stabilization) in addition to nitrogen, BOD, phosphorus, and suspended solids removal. A more detailed description of this combined nitrogen-phosphorus removal process is given in Chapter 8.

Design and operating data for the Palmetto, Fla., Bardenpho facility are shown in Tables 7.14 and 7.15.

PHYSICAL PROCESSES FOR PHOSPHORUS REMOVAL

This section of the manual deals with the use of physical processes for the separation of nutrient phosphorus from an aqueous stream. The title "Physical Processes" is somewhat of a misnomer in that in each case some chemical processes are involved. However, this definition has been applied in the present case to separate these physical-chemical processes from those operations or processes that involve only chemical precipitation. This section deals with three types of these processes. The first two are pressure-driven membrane processes and the third is the more familiar ion exchange method.

Ultrafiltration and reverse osmosis are both pressure-driven membrane processes, but they differ markedly in their mode of operation and in the nature of the materials that may be rejected by the membranes used. However, since both have in the past been used for removal of nutrients, they are properly included in this section.

Ion exchange processes are those processes in which an ion is selectively removed from a synthetic resin and replaced by an ion carried in the aqueous phase. While both natural and synthetic ion exchange materials exist, for present purposes, the discussion will be limited to the much more efficient synthetic ion exchange resins.

The physical processes have one comparatively undesirable feature in common. That is, they all produce a reject stream that must be treated specifi-

TABLE 7.14. Design conditions versus actual operating conditions (average values).

Parameter	Design	Actual
Daily flow, L/s (mgd)	61 (1.4)	44 (1.0)
Total detention time (hours)	11.6	16.2
Influent BOD, mg/L	270	135
Influent suspended solids, mg/L	250	134
Influent TKN, mg/L	43	32
Influent phosphorus, mg/L	14	7.0
MLSS, mg/L	3500	3346
Percent volatile suspended solids	—	70.0
SVI, mL/g	—	65
Temperature, °C	18–25	19–23
pH (nitrification zone)	—	6.8

TABLE 7.15. Palmetto, Fla., Bardenpho performance results (April 1981–March 1982). Monthly averages.

Influent	April	May	June	July	Aug.	Sept.	Oct.	Nov.	Dec.	Jan.	Feb.	March
Q, m³/d	3200	3000	3500	3600	5500	5900	3700	3300	3200	3600	3700	4500
Q, mgd	0.85	0.78	0.92	0.96	1.44	1.57	0.99	0.88	0.84	0.94	0.98	1.2
BOD$_5$ mg/L	164	159	124	104	74	67	113	157	182	160	163	150
SS, mg/L	155	157	144	112	76	76	116	160	182	141	167	128
Temperature, °C	25	27	29.5	30.5	29.5	29	28	27	24	23	23	23
TKN, mg/L	31.8	40.8	30.1	25.0	19.7	21.9	28.1	40.0	38.2	37.7	42.4	32.4
NH$_4^+$-N mg/L	25.0	25.2	20.4	18.7	12.7	12.7	17.8	22.6	27.2	28.0	25.8	23.7
Total P, mg/L	9.2	6.4	7.0	5.6	4.1	4.9	6.3	8.5	8.8	8.7	8.0	6.6
Ortho P, mg/L	6.5	6.1	5.3	4.5	2.8	3.5	4.7	5.9	5.9	5.4	5.2	4.4
Alkalinity, mg/L	174	169	156	154	143	140	144	171	198	191	201	187
Filtered effluent												
BOD$_5$ mg/L	2	1	1	1	1	1	1	1	1	1	1	1
SS, mg/L	3	2	2	2	2	1	2	2	2	1	2	3
Total N, mg/L	2.1	2.1	1.9	1.8	2.0	1.7	1.9	2.1	2.5	2.7	2.6	2.8
NO$_3$-N mg/L	1.0	1.3	1.0	1.9	1.2	1.1	1.1	1.3	1.5	1.9	1.8	2.1
NH$_4^+$-N mg/L	0.4	0.3	0.3	0.2	0.2	0.2	0.2	0.2	0.4	0.3	0.1	0.2
Total P, mg/L	2.5	3.4	2.6	1.8	1.5	1.2	1.1[a]	0.7[a]	1.6	0.6[a]	0.8[a]	0.9[a]
Ortho P, mg/L	2.2	1.4	2.5	1.7	1.1	1.3	0.9	0.7	1.0	0.5	0.7	0.8

[a] Minimum alum dosage applied.

cally and usually chemically for final removal of phosphorus. These processes should be viewed as methods of intercepting, but not necessarily finally containing or fixing the phosphorus contained in the aqueous stream.

Ultrafiltration systems. Ultrafiltration systems are those pressure-driven membrane operations that use porous membranes for the removal of dissolved and colloidal material from the aqueous phase. These systems usually are distinguishable from RO systems because of the relatively low driving pressures used, usually under 150 psi. Ultrafiltration is normally used to remove colloidal material and large molecules in true solution with molecular weights in excess of 5000. Ultrafiltration has been used in the past for removal of oil from aqueous streams, removal of turbidity in color colloids, removal of macro molecules in the pharmaceutical and fine chemical industry, for cleaning up paint-laden wastewaters, and other applications where comparatively large molecules or colloidal species are to be removed and where gravity sedimenta-

tion, following coagulation and flocculation, is not feasible.[50]

Ultrafiltration systems develop virtually no osmotic pressure differential across the membrane interface. This is so simply because the osmotic pressure is a function of the molarity of the system and the mole weight for the large molecules usually removed is a very large number, but the molarity of the system is low. As a consequence, obtaining an osmotic pressure difference from anything other than a fraction of molar strength is unlikely.

In actual practice, ultrafiltration systems usually use flows similar to that shown in Figure 7.8. In this case, the material to be subjected to ultrafiltration is received in a comparatively large receiving reservoir. The material is then pumped into the membrane system, which removes a comparatively small fraction of the waste introduced, with the remainder being recycled. Phosphorus that is contained as discrete suspended or colloidal material would be removed by UF. There is also a possible application for ultrafiltration in the removal of micro-colloids of calcium

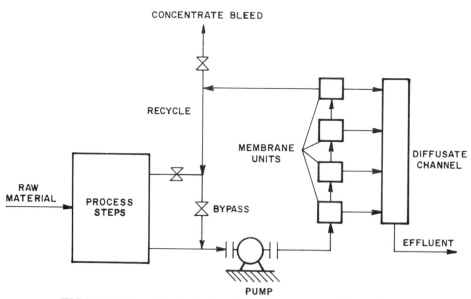

FIGURE 7.8. Typical ultrafiltration system flow sheet.

phosphate or the iron and aluminum salts of phosphates.

Reverse osmosis systems. Reverse osmosis differs from ultrafiltration in that the flow of water does not occur through discrete pores or holes in the membrane medium. It is thought that flow is facilitated by hydrogen bonding through interstices that occur between strands of the polymer that exist as it is laid down and cured. The curing process is employed to tighten one of the surfaces of film, hence, reverse osmosis membranes are anisotropic, that is, one surface is considerably different from the other and flow occurs in one direction. Because of this relative tightness, the pressure required to move water through most reverse osmosis membranes at zero osmotic pressure is on the order of 50 to 100 psi.

Membranes are normally made of such materials as cellulose diacetate and triacetate, polyamide, and polysulphone. The membranes are normally configured into bundles of tiny tubes often referred to as fibers, or the more conventional spiral wrapped configuration. Both the hollow fine fibers and the spiral wrapped configurations operate at such low velocities that pretreatment designed to remove colloids and gross suspended material is usually necessary for flux maintenance. The need for pretreatment can be appreciated when viewing the normal mode of pyramid construction and flow sheet of reverse osmosis systems (Figures 7.9 and 7.10). The normal pretreatment practice is included in Figure 7.11. As can be seen, the pretreatment may consist of coagulation, flocculation, sedimentation, and filtration for the removal of colloidal and gross solids. Normally, where the concentration of suspended solids in the feed stream does not exceed 30 to 50 mg/L, the clarification system is deleted. However, if there is not a convenient place for the discharge of the backwash water from the filter, a backwash clarifier normally is provided to conserve and reuse this portion of the spent water stream. The adequacy of the pretreatment system is frequently confirmed through the use of an index that quantifies the plugging or fouling, by percent, of the membrane. This index is sometimes called the Silting Index:

$$S.I. = \left(1 - \frac{T_2}{T_1}\right)(100) \qquad (5)$$

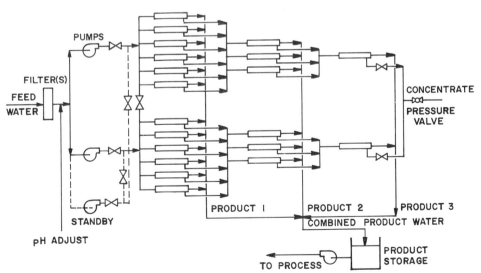

FIGURE 7.9. Pyramid construction for a single-stage reverse osmosis plant.

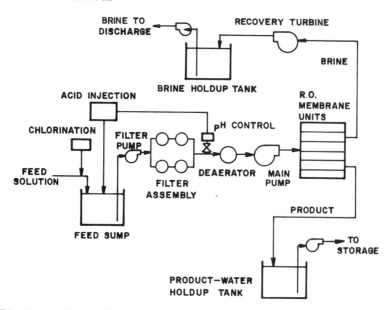

FIGURE 7.10. Basic flow diagram for a single-stage reverse osmosis plant.

where

  S.I. = the Silting Index,
  T_1 = is the time required to pass 500 mL through a 0.45-μm filter under a uniform head of 30 psi at time 0 and maintained throughout the test, and
  T_2 = the time required to pass 500 mL through the 0.45-μm membrane at a point 15 minutes after the initiation of the test and also under a uniform 30 psi head.

Values of the percent plugging or Silting Index under 30 are generally indicative of an extremely good water supply with a minimum of fouling tendency. Values of 30 to 60 generally are considered marginal, and values over 60 generally are indicative of an unsatisfactory water supply or one that will cause excessive fouling in the membrane system itself.

Membranes normally are rated as to their ability to remove sodium chloride. There are four broad classes of membranes normally marketed:

1. Seawater membranes that are capable of producing water in the neighborhood of 500 mg/L from seawater with a total dissolved solids (TDS) of approximately 30 000 to 40 000 mg/L;

2. 97 to 99% NaCl rejection membranes that normally are used with heavily brackish water (5000 to 15 000 mg/L TDS) to produce a high quality product, that is, one containing around 200 to 500 mg/L of TDS;

3. 95% NaCl rejection membranes, low salinity (less than 2000 mg/L); and

4. 90% NaCl rejection membranes, low salinity (less than 2000 mg/L).

The rejection efficiency must be based on the mean concentration of sodium chloride that the membrane "sees" in the pyramid system. This is roughly equal to the log mean and is a function of the extent of recovery of the raw water and the fraction that is rejected as brine.

If the recovery is very low, that is, on the order of 50%, there is only a one-fold increase in the concentration of salt across the membrane system. This permits the overall rejection to be estimated on the basis of the arithmetic average. If, on the other hand, the recovery is 80 or 90%, in which there is, respectively, a five-fold or a ten-fold increase in dissolved solids across the system, it

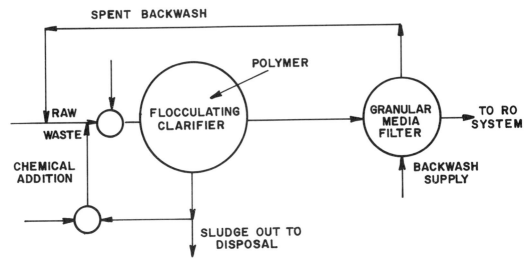

FIGURE 7.11. Reverse osmosis pretreatment system.

is much more important to use the actual concentration "seen" to estimate the actual rates using pilot or lab tests.

Phosphate is rejected very well, relative to sodium chloride. It can be expected, depending on the system and the other materials present, that between 95 and 99% of the phosphate seen by a 90% NaCl rejection membrane will be rejected. In the same fashion, 95 to 99% of the calcium seen by the same membrane will be rejected. This leads to increasing concentrations of species that tend to become insoluble, and thus requires that additional pretreatment be considered to assure that no precipitation of calcium carbonate, calcium sulfate, calcium fluoride, or calcium phosphate occurs. The order of rejection of components in treated effluent[51-53] by acetate and acetamide membranes are shown in Table 7.16.

Ion exchange systems. The heart of an ion exchange system is the resin that is capable of exchanging a specific ion on its surface for a specific ion or certain specific ions in the aqueous phase. The phenomenon of ion exchange was first noted in soils, for there are many natural materials with a substantial exchange capacity. Many of these, particularly the clay minerals, are referred to

as zeolites. This term is often applied to modern synthetic ion exchange materials. The synthetic resins now manufactured generally are of two types: those that exchange cations and those that exchange anions. Anion exchange resins are of interest here. Anion exchange capacity is brought about by the insertion of either quaternary amine or an amine group in a condensation reaction producing a long chain polymer, which is made into a resin bead. These beads have substantial permeability and not only can the material on the surface be exchanged, but as the aqueous phase passes through, the exchange sites on the interior also are used. The exchange capacity of most anionic resins is on the order of 4 to 5 meq/g.

TABLE 7.16. Order of rejection of components in treated effluent cellulose acetate blend and acetamide membranes.

Component	% Rejection	
	Ref.[51]	Ref.[52]
PO$_4$	87–99	93–99
NO$_3$	33–95	33–97
NH$_4$	62–96	
COD	42–99	68–97
Cl	47–78	

Ion exchange resins usually are packed in columns. As a consequence, there is virtually no tolerance for the presence of suspended or colloidal material, which after contact, can settle or adhere to the surface of the resin itself. As a consequence, prior to ion exchange, substantial pretreatment is required for the removal of colloids and gross suspended solids. This is normally accomplished in the same way that such pretreatment is carried out for reverse osmosis. That is, if the suspended solids are comparatively high, and the system contains colloids as well as suspended material, coagulation, flocculation, and sedimentation, followed by granular media filtration are used. If the raw water contains less than 50 mg/L, the sedimentation step is often deleted and the chemically conditioned raw water is fed directly to the granular media filter.

Exchangeability and selectivity in resins are related to the nature of the group exchanging the ion, the valence of the ion going onto the resin, and the ratio of the activity of that ion inside and outside the resin. Much of the basic thermodynamics used to define chemical equilibria can be used in the chemistry of the ion/resin combination.

At equal concentration, trivalent materials generally will be exchanged preferentially over divalent materials, which in turn will be exchanged preferentially over monovalent materials. However, because of the comparatively low concentration of phosphate in most effluents, phosphate would not be removed selectively by most of the resins available. There are, however, some resins that are selective for phosphorus of concentrations higher than normally are seen in most effluents. It would be in

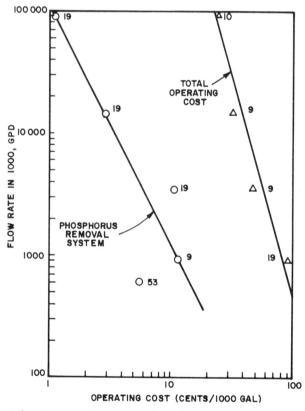

FIGURE 7.12. Plant operating costs for chemical removal of phosphorus (1979).
NOTE: 1000 gpd $\times$ 3.785 = m^3/d; ¢/1000 gal $\times$ 3.785 = ¢/m^3.

TABLE 7.17. Capital and operating costs for chemical phosphorus removal systems.

| Capacity | | Capital costs ($)[a] | | Non-chemical operating costs, ¢/m³ (¢/1000 gal)[a] | |
L/s	mgd	1975	1983	1975	1983
44	1	10 000	22 000	.13 (0.50)	0.26 (1.00)
220	5	30 000	66 000	.05 (0.18)	0.09 (0.35)
440	10	53 000	117 000	.04 (0.15)	0.08 (0.30)
880	20	95 000	210 000	.03 (0.13)	0.07 (0.26)

[a] Chemical addition equipment installed; Fe^{+3}/Al^{+3} and polymer.

these particular cases that ion exchange would have its best application.

Most of the resins that would be employed for the removal of phosphate or mono or dihydrogen phosphate, would be basic in character and require the use of either a strong base or a moderately strong base, such as sodium carbonate, for regeneration. The disposal of the regeneration materials—both the wash, strong regenerant, and the rinse waters—often poses a special problem, unless these regenerant streams can be introduced conveniently into some point nearby in an in-place wastewater treatment facility. In addition, it will be necessary to have a secondary or supporting sedimentation facility for the removal of the phosphate as either the calcium iron or aluminum salt.

COSTS OF PHOSPHORUS REMOVAL

Phosphorus removal by chemical procedures. The costs of chemical procedures are limited to the operating costs associated with the addition of the chemical species. Chemical addition systems and mixing systems are comparatively inexpensive. The operating costs for chemical precipitation systems are shown in Figure 7.12[9,10,19,54] and Tables 7.17 and 7.18.[54] These values are to be used only as guides.

Phosphorus removal by physical procedures. Included in this section are the costs of operating a reverse osmosis system and an ion exchange system. The reverse osmosis system is operated at 80% recovery. The estimated capital and operating costs for these facilities

TABLE 7.18. Basic statistics from the Canada–Ontario full-scale treatability study program.

| | | Addition to raw wastewater | | | Addition to mixed liquor | | |
Chemical	Total no. of plants	No. of plants	Avg. dose (mg/L)	Avg. cost[c], $/ML ($/million gallons)	No. of plants	Avg. dose (mg/L)	Avg. cost[c], $/ML ($/million gallons)
Iron	22	7	16.0[a]	10.6 (40.3)	15	9.2	5.9 (22.5)
Alum	20	5	10.3[a]	17.7 (67.0)	15	7.2	11.5 (43.4)
Lime	3	3	118[b]	9.2 (35.0)	—	—	—

[a] Dosages as mg/L Fe/Al.
[b] Dosages as mg/L Ca.
[c] Chemical cost only (adjusted December 1979). The 1979 operating costs may be adjusted in the approximate fashion to 1983 by multiplying by 1.4. However, local circumstances, particularly with respect to chemical costs, will have a greater bearing on increasing or changing costs than the overall index.
The 1983 cost breakdown essentially is as follows:
Alum $0.13-$0.22/kg ($0.06-$0.10/lb) Labor   20%–30% of total cost
FeCl₃ $0.02-$0.06/kg ($0.01-$0.03/lb) Chemical 70%–80% of total cost (Fe or Al)
Lime $0.02-$0.09/kg ($0.01-$0.04/lb) Labor   40%–65% of total cost
Polymer $3.30-$6.60/kg ($1.50-$3.00/lb) Chemical 35%–50% total cost (Lime)

163

TABLE 7.19. Physical phosphorus removal systems; reverse osmosis and ion exchange; capital and operating costs. (Salinity less than 1000 mg/L; pressure less than 500 psi).[a]

Plant Size		Capital costs ($)	Operating cost, ¢/m³ (¢/1000 gal)
L/s	mgd		
		Reverse osmosis	
4	0.1	200 000	30 (115)
9	0.2	300 000	24 (90)
22	0.5	400 000	20 (75)
44	1.0	750 000	17 (65)
220	5.0	2 000 000	15 (55)
440	10.0	3 000 000	13 (50)
		Ion exchange	
4	0.1	160 000	33–92 (125–350)
22	0.5	230 000	29–85 (110–320)
44	1.0	300 000	26–193 (100–300)

[a] Power—4¢/kWh; acid—¢6/kg (¢3/lb); membrane life—3 years; based on cost data provided by Roga, Rohm & Haas and Dow, Inc.
Note: psi × 6895 = Pa.

are shown in Table 7.19. Similar costs for ultrafiltration are not included because of the experimental nature of the process.

REFERENCES

1. De Pinto, J.V., et al., "Phosphorus Removal in Lower Great Lakes Municipal Treatment Plants." In "International Seminar on Control of Nutrients in Municipal Wastewater Effluents." Vol. I, U.S. Environ. Prot. Agency, MERL, Cincinnati, Ohio (Sept. 1980).
2. Stumm, W., Discussion in "Advances in Water Pollution Control Research." Proc. 1st Int. Conf. Water Pollut. Res., Pergamon Press, Ltd., London, 2, 216 (1964).
3. Leckie, J., and Stumm, W., "Phosphate Precipitation in Water Quality Improvement by Physical and Chemical Processes." Univ. of Texas, Austin, 237 (1970).
4. Stam, H., and Kohlschuetter, H. W., "Sorption of Phosphate Ions on Iron Hydroxide." J. Inorg. Nuclear Chem., 27, 2103 (1965).
5. Shannon, E., "Physical-Chemical Phosphorus Removal Processes." Paper presented at Nutrient Control Seminar, Calgary, Alberta, Canada (1980).
6. Nesbitt, J., "Phosphorus Removal—The State of the Art." J. Water Pollut. Control Fed., 41, Part 2, 5, 701 (1969).
7. Grutsch, J., and Mallatt, R., "Optimize the Effluent System." Hydrocarbon Processing, 56, 3-7 (1976).
8. Ketchum, B. H., "The Absorption of Phosphate and Nitrate by Illuminated Cultures of Nitzschia Closterium." Am. J. Botany, 26, 399 (1939).
9. von Smoluchowski, M., "Versuch einer mathematischen Theorie der Koagulatinnkinetik Kolloider Losungen." Zeit. fur physikalische Chemie, 92, 155 (1917).
10. EIMCO PEC, unpublished internal reports on recycle effects (1979-1983).
11. The Dow Company, "Phosphorus Removal From Wastewater By Chemical Precipitation and Flocculation." Midland, Mich. (1972).
12. Sutton, P., et al., "Reliability of Nitrification Systems With Integrated Phosphorus Precipitation." Rept. 75-3-21, Ont. Ministry of Environ., Toronto, Canada (1975).
13. Schmidtke, N. W., "Sludge Generation, Handling and Disposal at Phosphorus Control Facilities." In "Phosphorus Management Strategies for Lakes." Ann Arbor Science, Ann Arbor, Mich. (1980).
14. Schmidtke, N. W., "Nutrient Removal Technology—The Canadian Connection." In "International Seminar on Control of Nutrients in Municipal Wastewater Effluents." 1, "Phosphorus, Vol. 1" U.S. Environ. Prot. Agency, Cincinnati, Ohio (1980).
15. Allied Chemical Co., Wastewater News.
16. Allied Chemical Co., Wastewater News (March 1976).
17. Allied Chemical Co., Wastewater News (Jan. (1973).
18. Allied Chemical Co., Wastewater News (Aug. 1973).
19. Switzenbaum, M., et al., "Phosphorus Removal: Field Analysis." Proc. J. Am. Soc. Civ. Eng., 107, 1171, (1981).

20. Rugby, L. F., "Tertiary Chemical Treatment for Phosphate Removal." Paper presented at 43rd Annual Meeting, Ohio Water Pollut. Control Conf., Toledo, Ohio (1969).

21. Stepko, W. E., and Schroeder, W. H., "Design Consideration to Attain Less Than 0.3 mg/L Effluent Phosphorus." Canadian Ontario Agreement Conf. Proc. No. 3, High Quality Effluent Seminar, Toronto, Canada, 179 (Dec. 1975).

22. Rudolph, W., "Phosphates in Sewage and Sludge Treatment, II. Effect on Coagulation, Clarification and Sludge Volume." *Sew. Works J.,* **19,** 179 (1947).

23. Sawyer, C.N., "Some New Aspects of Phosphates in Relation to Lake Fertilization." *Sew. Ind. Wastes,* **24,** 768 (1952).

24. Owen, R., "Removal of Phosphorus From Sewage Plant Effluent With Lime." *Sew. Ind. Wastes,* **24,** 768 (1953).

25. Menar, A., and Jenkins, O., "The Fate of Phosphorus in Waste Treatment: The Enhanced Removal of Phosphate By Activated Sludge." Proc. of 24th Purdue Ind. Waste Conf., Purdue Univ., Lafayette, Ind. (1969).

26. Wuhrman, K., "Objective, Technology and Results of Nitrogen and Phosphorus Removal Processes." In "Advances in Water Quality Improvement." Univ. of Texas Press, Austin (1968).

27. Spiegel, M., and Forrest, T., "Phosphate Removal: Summary of Papers." *Proc., J. Am. Soc. Civ. Eng.,* **95,** 803 (1969).

28. Albertson, O., and Sherwood, R., "Phosphate Extraction Process (PEP)." *J. Water Pollut. Control Fed.,* **41,** 1467 (1969).

29. Schmid, L., and McKinney, R., "Phosphate Removal by a Lime-Biological Treatment Scheme." *J. Water Pollut. Control Fed.,* **41,** 1259 (1969).

30. Buzzell, J., and Sawyer, A., "Removal of Algal Nutrients from Raw Wastewater With Lime." *J. Water Pollut. Control Fed.,* **39,** R16 (1967).

31. Van Wazer, J., "Phosphorus And Its Compounds." Interscience Publ., Inc., New York, N.Y., **1** (1958).

32. Clark, J., "Solubility Criteria For the Existence of Hydroxyapatite." *Can. J. Chem.,* **33,** 1696 (1955).

33. Bogan, R., *et al.,* "Use of Algae in Removing Phosphorus from Sewage." *Proc. J. Am. Soc. Civ. Eng.,* **86,** (1960).

34. Jebens, H., and Boyle, W., "Enhanced Phosphorus Removal in Trickling Filters." *Proc. J. Am. Soc. Civ. Eng.,* **98,** 547 (1972).

35. Humenick, M., and Kaufman, W., "An Integrated Biological Chemical Process for Municipal Wastewater Treatment." Paper presented at 5th Int. Water Pollut. Res. Conf., San Francisco, Calif. (1970).

36. Ferguson, J., *et al.,* "Calcium Phosphate Precipitation in Wastewater Treatment." In "Water, 1970." Am. Inst. Chem. Eng.(1970).

37. Menar, A., and Jenkins, D., "Calcium Phosphate Precipitation in Wastewater Treatment." EPA-R2-72-064, U.S. EPA(1972).

38. Davis, S., "Alternatives for Phosphate Removal." *Water Sew. Works* (1970).

39. Bishop, D., *et al.,* "Physical Chemical Treatment of Municipal Wastewater." *J. Water Pollut. Control Fed.,* **44,** 361 (1972).

40. "Advances in Wastewater Treatment, Pilot Plant, Pomona, California." FWQA and L. A. County San. Dis., Los Angeles, Calif. (1969).

41. Schulze, K. L., "Tertiary Treatment in Wastewater." *Deu Ind Micro,* **9,** 179 (1968).

42. Albertson, O. E., Private Communication (1982).

43. Eberhardt, W., and Nesbitt, J., "Chemical Precipitation of Phosphorus in a High Rate Activated Sludge." *J. Water Pollut. Control Fed.,* **40,** 1239 (1968).

44. EIMCO PEC, unpublished reports on sedimentation loading rates in water and waste treatment (1978-1983).

45. Greenberg A. E., *et al.,* "Effect of Phosphorus Removal on Deactivated Sludge Process." *Sew. and Ind. Waste,* **27,** 277 (1955).

46. Srinath E. J., *et al.,* "Rapid Removal of Phosphorus from Sewage by Activated Sludge." *Experientia (Switzerland),* **15,** 339 (1959).

47. Levin G. V., and Shapico, J., "Metabolic Uptake of Phosphorus by Wastewater Organisms." *J. Water Pollut. Control Fed.,* **37,** 800 (1965).

48. Bernard, J. L., "Cut P and N Without Chemicals." *Water and Waste Eng.,* **7,** (1974).

49. Masse, D. L., "The PhoStrip Process for Biological Removal of Phosphorus from Wastewater." Pres. Workshop on Biological Phosphorus Removal in Municipal Wastewater Treatment." Annapolis, Md. (June 1982).

50. "New Technology For Treatment of Wastewater by Reverse Osmosis." Envirogenics Co., El Monte, Calif., EPA-17020DUD 09/70 (1970).

51. Boen, D., and Johannsen, G., "Reverse Osmosis of Treated and Untreated Secondary Sewage Effluent." EPA-670/2-74-077 ORD, U.S. EPA, Cincinnati, Ohio (1974).

52. Okey, R., *et al.,* "Engineered Membrane Systems, A New Dimension in Waste Management." Paper presented to Am. Chem. Soc. Annu. Meeting, Miami Beach, Fla. (1967).

53. Archer, J., "Summary Report on Phosphorus Removal." Canada—Ontario Agreement Research Report #83 (1978).

54. Shannon, E., "Physical-Chemical Phosphorus Removal Processes." Paper presented at Nutrient Control Seminar, Calgary, Alberta, Canada (1980).

Chapter 8
Combined Phosphorus and Nitrogen Removal

The purpose of this chapter is to describe those phosphorus and nitrogen removal systems that were designed specifically to remove either one or the other nutrient but that also, without major changes in chemical addition or process modification, were found to provide removal of both.

BACKGROUND

In general, the two categories of approach to developing a treated wastewater are physical/chemical treatment and biological treatment. The essential difference between physical/chemical processes and biological processes is the ability of each to remove certain materials. Although both categories suffer from inefficiencies, each is highly efficient in its proper application.

There are situations, however, when a process, with some modifications, can be used to treat several waste constituents. Once the total strength and the soluble and particulate portions of a waste have been determined, the anticipated performances of biological and physical/chemical treatment systems can be estimated. Table 8.1[1] shows typical process capabilities for the treatment of domestic wastewater.

The systems described below are modifications of the suspended growth unit processes shown in Table 8.1 that

have demonstrated some potential for removal of both nitrogen and phosphorus. They include the Bardenpho Process®, the A/O Process®, and the Phostrip Process®.

Bardenpho Process. As described in Chapter 7, the Bardenpho Process is a modification of the activated sludge process. The process was developed to remove 80 to 90% of all nitrogen in a domestic wastewater through a series of anoxic and aerobic chambers (Figure 8.1).

During bench- and pilot-scale tests with the Bardenpho Process, it was found that good phosphorus removal would also take place at times, with an apparent correlation between the presence of nitrates in the effluent and the inability of the plant to remove phosphorus (Figure 8.2). Experimentation showed that for the removal of phosphorus to succeed, the bacteria must pass through a short period during which fermentation takes place. To ensure that the fermentation stage took place, a small basin was added ahead of the first anoxic basin. The underflow from the clarifier is returned to this first fermentation basin, as well as some of the raw or settled wastewater, while mixed liquor from the aeration basin is recycled to the first anoxic basin. Figure 8.3 shows the Bardenpho Process configuration with the addition of a fermentation zone.

TABLE 8.1. Various combinations of biological and physical–chemical unit processes.

Raw waste constituent	Typical concentration mg/L				First unit		Second unit		Third unit	
	Soluble	Colloidal	Solid	Total	Unit process	Typical out concentration (mg/L)	Unit process	Typical out concentration (mg/L)	Unit process	Typical out concentration (mg/L)
Suspended solids	—	—	200	200	sedimentation	80–100	act. sl.	10–30	filtration	3–7
					coag./sed.	10–30	act. sl.	10–30	filtration	3–7
					coag./sed.	10–30	filtrat.	3–7		
					act. sl.	10–30	filtrat.	3–7		
BOD$_5$ (carbonaceous)	80	40	80	200	sedimentation	130–150	act. sl.	10–30	filtration	1–3
					coag./sed.	80–100	act. sl.	10–30	filtration	1–3
					coag./sed.	80–100	filtrat.	80–90	adsorption	5–15[a]
					act. sl.	10–30	filtrat.	1–3	adsorption	0–2
COD	160	80	160	400	sedimentation	240–300	act. sl.	50–100	filtration	40–60
					coag./sed.	160–200	act. sl.	50–100	filtration	40–60
					coag./sed.	160–200	filtrat.	160–180	adsorption	20–30
					act. sl.	50–100	filtrat.	40–60	adsorption	5–10
Phosphorus	9	—	1	10	coag./sed.	2–5	filtrat.	0–1	filtration	0–2
					coag./sed.	2–5	act. sl.	1–5		
Nitrogen, total, as N				20–85	sedimentation	18–80	filtrat.	16–76	filtration	10–76
							act. sl.	18–80	filtration	16–76
									adsorption	16–76

[a] No credit shown for removal caused by biological activity.

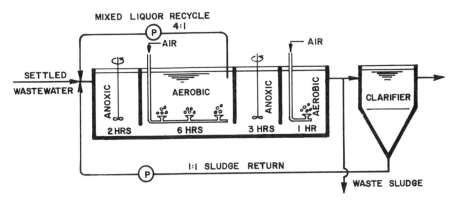

FIGURE 8.1. Diagram of Bardenpho Process.

As a result of the research and pilot testing performed by J. C. Barnard and others[2], a number of pilot plants and full-scale facilities were constructed in South Africa. Phosphorus removals in the final effluents of activated sludge plants have been achieved to a level between 0.2 and 0.8 mg/L, together with removal of nitrogen of between 80 to 90%.[2,3]

The first full-scale Bardenpho wastewater treatment facility in the U.S. has been operating at Palmetto, Fla., since October 1979 (Figure 8.4). The system design of the plant is shown in Table 8.2. Operating conditions are shown in Table 8.3. Average operating performance are shown in Tables 8.4 and 8.5.

The results of a study conducted during the first few months of operation of the facility showed that BOD_5, suspended solids (SS), total nitrogen, and phosphorus averaged less than 3.0, 5.0, 2.5, and 3.0 mg/L, respectively. The conclusions drawn from the work done in Florida[3] are that the Bardenpho system must be designed specifically for each application as a function of wastewater influent concentrations of BOD_5, SS, nitrogen and phosphorus, and wastewater temperature. For weak wastewaters or wastewaters with relatively high phosphorus concentrations, some use of chemicals for effluent polishing may be necessary to reduce phosphorus concentration to less than 1 mg/L.

A/O Process. As described in Chapter 7, the A/O Process functions by subjecting activated sludge microorganisms to alternating anaerobic/aerobic conditions. Initially, laboratory and pilot studies were conducted to study the removal of phosphorus through manipulation of the activated sludge process. The studies showed that phosphorus could be removed through proper control of the process variables along with chemi-

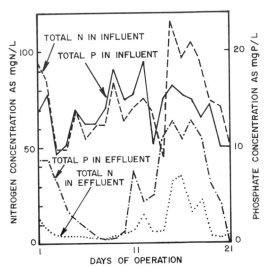

FIGURE 8.2. Correlation between nitrates and phosphorus in the effluent.

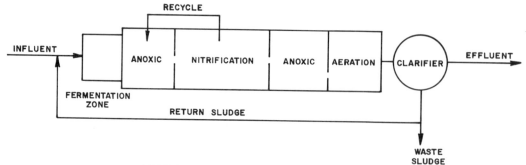

FIGURE 8.3. Bardenpho schematic.

cal stripping of phosphorus from the recycle sludge. Later observations showed that phosphorus removal was being achieved in several full-scale plants even though these plants were not designed for phosphorus removal.

The results of laboratory and pilot studies combined with plant observations yielded some features common to those plant configurations where phosphorus was being removed. Figure 8.5 shows a schematic representation of the A/O Process.

As the desired configuration that now is known as the A/O Process became better understood, researchers were able to experiment with changes and additions that could also effect nitrogen

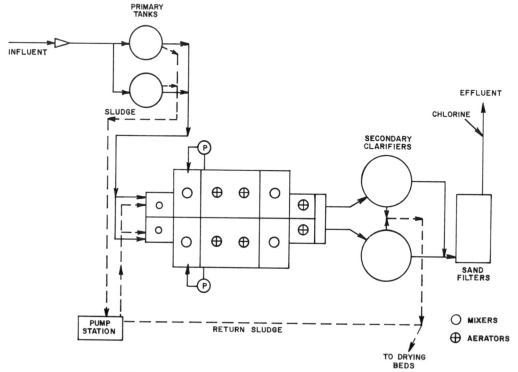

FIGURE 8.4. Bardenpho facility. Palmetto, Fla.

TABLE 8.2. Bardenpho system design. Palmetto, Fla.

Parameter	Fermentation	First anoxic zone	Aeration zone	Second anoxic zone	Reaeration zone
Detention time (hours)	1.0	2.7	4.7	2.2	1.0
Mixing or aeration hp per train	1-5	1-5	2-20 Two-speed	1-5	1-5
Internal recycle	2/Train — Each 4 × Influent flow			37 kW (50-hp) Blower used for aeration	

Design	
Flow	1.4 mgd (61 L/s)
BOD$_5$	270 mg/L
Total solids retention time	20 days

removal. As a result of laboratory experiments and additional pilot studies, a 400 L/s (9-mgd) plant in Largo, Fla., was used as a demonstration plant and a full-scale research and development unit to demonstrate nutrient control.[4] The city's wastewater treatment plant consists of three identical contact stabilization activated sludge trains. Each train has the configuration shown in Figure 8.6 and handles approximately 130 L/s (3 mgd).

During the course of the 12-month Largo program, the A/O system was operated in both the non-nitrifying and nitrifying modes. Operating conditions are shown in Tables 8.6 and 8.7. The system as set up for nitrification and denitrification is shown in Figure 8.7. Basin layout is shown in Figure 8.8. Daily grab samples from the influent and from various locations throughout the process showed nitrogen removals (as NO$_x$—N) in the effluent approximately 66%

TABLE 8.3. Design conditions versus actual operating conditions (average values).

Parameter	Design	Actual	Actual fraction of design
Daily flow, L/s (mgd)	61 (1.4)	53 (1.22)	38 (0.87)
Total detention time (hr)	11.6	13.3	1.15
Influent BOD$_5$ (mg/L)	270	164	0.60
Influent suspended solids (mg/L)	250	143	0.57
Influent TKN (mg/L)	43	32	0.75
Influent phosphorus (mg/L)	14	8.4	0.60
MLSS (mg/L)	3500	3346	0.98
Percent volatile suspended solids	—	70.0	—
SVI (mL/g)	—	65	
Temperature	18°-25°C	19°-23°C	
pH (nitrification zone)		6.8	

TABLE 8.4. Bardenpho average operating performance; BOD₅, suspended solids, nitrogen.

	BOD$_5$		Suspended solids		Total nitrogen (TKN + NO$_3$-N)		TKN		NO$_3$-N
	mg/L	% removal	mg/L	% removal	mg/L	% removal	mg/L	% removal	mg/L
Influent	164.0		143.0		32.0		32.0		
Secondary clarifier effluent	2.5	98.5	4.8	96.6	2.3	92.8	1.0	96.9	1.3
Sand filter effluent	1.7	98.9	2.0	98.6	2.1	93.4	0.8	97.5	1.3

TABLE 8.5. Bardenpho average operating performance; phosphorus removal.

| | Total phosphorus | | | | | | | |
| | A (1/1-1/10) | | B (1/11-2/7) | | C (2/7-3/10) | | D (3/20-4/20) | |
Phase[a]	mg/L	% removal	mg/L	% removal	mg/L	% removal	mg/L	% removal
Influent	9.0		8.3		7.9		8.4	
Secondary clarifier effluent	4.5	50	3.4	59	2.3	71	2.9	65
Sand filter effluent	4.5	50	3.4	59	2.2	72	2.8	67

[a] Date.

(Figure 8.9). Unfortunately, no data were given describing phosphorus removal from the A/O mode with nitrification. Phosphorus effluent levels averaged 1 mg/L (as P) for the A/O mode without nitrification–denitrification (Figure 8.10).

The Phostrip Process. The removal of phosphorus using a combination of biological and chemical treatment is the basis of the Phostrip Process. Described fully in Chapter 7, the system takes advantage of the activated sludge unit process by inducing the dominant bacterial solids to remove phosphorus from the wastewater and then to release it in a sidestream in the form of orthophosphate. The second step consists of precipitating the phosphate from this sidestream by means of lime addition (Figure 8.11).

It is possible to modify the Phostrip Process and thus accomplish removal of both nitrogen and phosphorus. A description of the nitrification–denitrification modification to the conventional activated sludge has been given in Chapters 5 and 6. As was discussed in Chapter 6, the key to the nitrification process design is to ensure that the sludge retention time is long enough so that the nitrifying bacteria are maintained at a sufficiently high concentration to be able to accomplish the oxidation of ammonia to nitrate. This can be achieved in the Phostrip system by designing for a sufficiently long sludge retention time in the existing anoxic tank

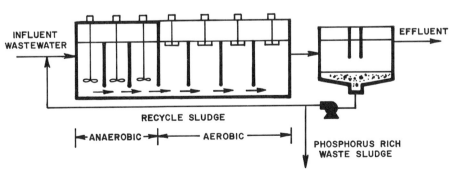

FIGURE 8.5. The A/O system for BOD and P removal with nitrification and denitrification.

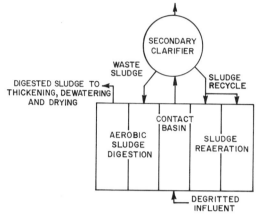

FIGURE 8.6. Contact stabilization basin at city of Largo, Fla.

TABLE 8.6. Average operating conditions and performance during A/O operation for removal of BOD and P.

Operating conditions	Period 1	Period 2
Influent flow,		
L/s (mgd)	170 (3.9)	140 (3.2)
Basin temperature, °C	24	25
MLSS, mg/L	3300	2300
IDT, hours		
Anaerobic/oxic	1.2/2.1	1.5/2.5
Performance	**Effluent**	**Effluent**
Total BOD, mg/L	9.8	6.4
Soluble BOD, mg/L	3.2	2.8
Total P, mg/L	—	1.2
Soluble P, mg/L	0.56	0.5
TSS, mg/L	20	11

(the "stripper" tank in Figure 8.12). The underflow from the clarifier is conducted to a denitrification tank where the sludge is exposed to anoxic conditions for about 2 hours. The effluent sludge is then split. About half goes back to the aeration tank and the other half to the stripper tank where orthophosphates are released.

Although a denitrification tank is required for the Phostrip system to achieve nitrogen as well as phosphorus removal (Figure 8.12), the system is not dependent on chemical additives (organic sources such as methanol) to achieve denitrification; rather, the organic impurities in the wastewater influent are used as a carbon source. It is claimed that 70% nitrogen removal is achieved when this process modification is used.[5] Total phosphorus removals range from 88 to 99% (Chapter 7).

TABLE 8.7. Operating conditions for full-scale Largo operation in the nitrifying A/O modes.

Operating parameter	June 5-July 31 A/O with nitrification	March 2-24 A/O with nitrification and denitrification
Influent flow, L/s (mgd)	140 (3.24)	140 (3.16)
No. of anaerobic/anoxic/oxic stages	5/—/5	3/2/5
Internal recycle flow, L/s (mgd)	—	160 (3.64)
Oxic section average pH	6.88	6.83
Mixed liquor temperature, °C	28.6	23.6
Oxic section IDT, hours	2.51	2.58
Oxic section NRT, hours	1.89	1.09
Oxic section SRT, d	4.0	3.7
F_s/M, kg/kg·d (lb/d/lb MLVSS)	0.10	0.13
MLSS, mg/L	3200	3450

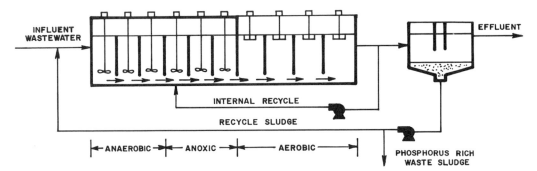

FIGURE 8.7. Flow diagram of the A/O system.

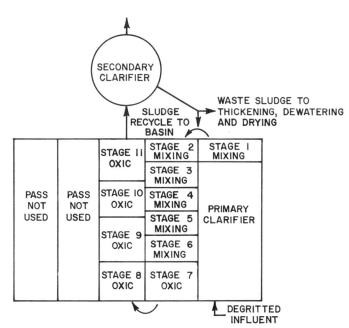

FIGURE 8.8. Basin layout of a full-scale A/O system. Largo, Fla.

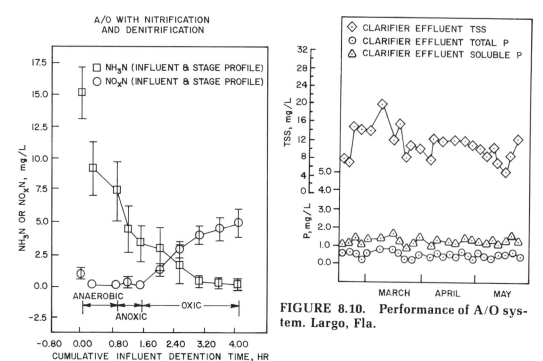

FIGURE 8.9. Mean data values with standard deviations for nitrogen removal in a full-scale A/O system. Largo, Fla.

FIGURE 8.10. Performance of A/O system. Largo, Fla.

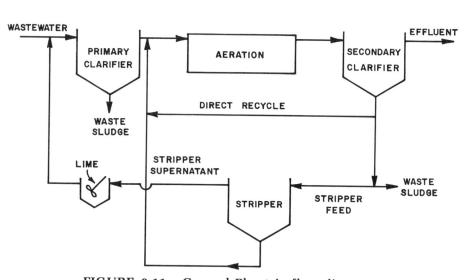

FIGURE 8.11. General Phostrip flow diagram.

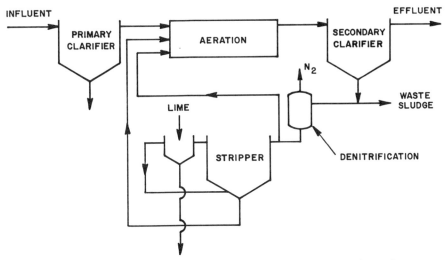

FIGURE 8.12. **Combined phosphorus-nitrogen removal. Phostrip Process.**

SUMMARY

While there appears to be successful use of biological phosphorus removal processes, more data are required to demonstrate fully the processes and to define limitations. The design engineer should update performance data prior to finalizing a design.

REFERENCES

1. "Wastewater Treatment Plant Design." Manual of Practice No. 8, Water Pollut. Control Fed., Washington, D.C.; Manual of Engineering Practice No. 36, Am. Soc. Civ. Eng., New York, N.Y. (1977).

2. Barnard, J. L., "The Bardenpho Process." In "Advances in Water and Wastewater Treatment Biological Nutrient Removal." Ann Arbor Science Publ. Inc., Ann Arbor, Mich. (1978).

3. Stensel, H. D., et al., "Performance of First U.S. Full-Scale Bardenpho Facility." In "International Seminar on Control of Nutrients in Municipal Wastewater Effluents. Proc. Vol. III: Nitrogen and Phosphorus." U.S. Environ. Prot. Agency, MERL, Cincinatti, Ohio (1980).

4. Hong, S., et al., "A Biological Wastewater Treatment System for Nutrient Removal." Presented at the 54th Annual Conference Water Pollut. Control Fed., Detroit, Mich. (1981).

5. Matsch, L. C., and Drnevich, R. F., "Phostrip: A Biological-Chemical System for Removing Phosphorus." In "Advances in Water and Wastewater Treatment. Biological Nutrient Removal." M. P. Wanielista and W. W. Eckenfelder, Jr. (Eds.), Ann Arbor Science Publ. Inc., Ann Arbor, Mich. (1978).

Chapter 9

Land Treatment

Land treatment, the practice of applying municipal and industrial wastewater and treated effluent to land, has received considerable attention as a means of nutrient removal. Previously, the objectives were either wastewater disposal or water reuse. But more recently, an increasingly important goal has been the additional treatment of wastewater to improve stream quality or efficient use of nutrients in wastes for raising crops. Under controlled application rates with aerobic conditions within the soil, plant nutrients and other residual substances are removed and degraded by: microorganisms in the surface soil horizons, chemical precipitation, ion exchange, biological transformation, and biological uptake through the root systems of the vegetative cover.

Because the water to be returned to the soil system is effluent from domestic and industrial uses, it is important to understand the ability of the soil system to alter water quality, and to minimize long-term adverse effects. Preliminary investigations for the purpose of establishing the general feasibility of a land application site can be conducted at minimum cost without extensive laboratory field work, and in some circumstances by inspection only if the combined suitability of soil, water, topography, and location is evaluated by experts. In general, the issues of greatest importance are to prevent groundwater pollution by achieving optimum quality improvements by the soil microorganisms and plants; and to achieve maximum hydraulic application rates consistent with the protection of both soil and crops and the elimination of runoff into streams.[1]

The process of wastewater purification and treatment by land is complex because of the large number of interacting variables involved. While this complexity influences achievement of optimum design, it does not interfere with present day capabilities to design and implement successful engineering systems. This chapter concentrates principally on the degree of wastewater renovation for nutrient removal under given water-soil-crop conditions.

There are three types of land treatment processes: slow rate (irrigation), rapid infiltration, and overland flow. All are capable of nutrient removal. Slow rate land treatment is the application of wastewater to vegetated soils usually for producing a salable crop, as well as for treating wastewater. Typically, 50 to 70% of the applied wastewater goes to plant uptake and evaporation, and the remainder percolates down through the soil. Loading rates range from 0.6 to 6 m/a (2 to 20 ft/yr), but are typically from 1 to 2 m/a (3 to 6 ft/yr).

Rapid infiltration involves intermittent flooding of shallow basins with relatively coarse textured soils. Nearly all of the applied wastewater percolates through the soil profile. Loading rates range from 6 to 180 m/a (20 to 600 ft/yr), but are typically from 15 to 60 m/a (50 to 200 ft/yr).

Overland flow is the application of wastewater to a smooth vegetated slope of slowly permeable soil. Applied wastewater is treated as it moves in a thin film over the soil and down the slope to collection ditches. Typically, 50 to 60% of the applied wastewater runs off (treated effluent); 30% evaporates; and 10 to 20% percolates into the soil. Loading rates range from 3 to 21 m/a (10 to 70 ft/yr), but are typically from 6 to 12 m/a (20 to 40 ft/yr).

The mechanisms and treatment efficiencies for nitrification and removal of nitrogen, BOD, phosphorus, and micronutrients from applied wastewater are described in this chapter. Groundwater protection and costs are also discussed.

NITRIFICATION IN THE SOIL ENVIRONMENT

When municipal wastewater is applied to land, the predominant form of nitrogen in wastewater is usually ammonium, although some nitrate is also likely to be present if the preapplication treatment processes have included one

or more aerobic stages. A small quantity of organic nitrogen, of which a part is soluble and readily convertible to ammonium through microbial action, is also usually present. Insoluble organic nitrogen associated with the particulate matter is also convertible to ammonium, although somewhat more slowly. Thus, for all practical purposes, the nitrogen form existing in domestic wastewater is ammonium nitrogen and concentrations typically range from 15 to 50 mg/L N for untreated domestic wastewater and 20 mg/L N for biologically treated effluent before land application.

The chemistry of nitrogen in the soil environment is complex because of the several oxidation states that nitrogen can assume and because changes in oxidation state can be brought about by the combined effects of chemical and microbial action. Plant uptake of nitrogen, volatilization of nitrogen through biological or chemical denitrification, escape of ammonia (NH_3) from alkaline soils, and nitrate (NO_3) leaching are the main mechanisms for nitrogen removal from soil solution caused by the application of wastewater on land.

Volatilization and adsorption of ammonia. The chemical equilibrium between the molecular (free) ammonia and the ammonium ion is pH-dependent, as shown in the following equations (Equations 1 and 2):

$$NH_3 + H^+ \rightleftharpoons NH_4^+ \qquad (1)$$

and

$$pH = 9.3 - \log \frac{[NH_4^+]}{[NH_3]} \qquad (2)$$

The proportion of molecular ammonia is small at pH values below 8, which are common in most domestic wastewater. Volatilization can be expected to be less than 10% between pH 7 and 8. Above pH 7.8, the potential for volatilization increases significantly.

A preliminary step to nitrification is adsorption of ammonium onto the negatively charged clay and organic colloids in the soil. In slow rate systems, the ammonium adsorption capacity of soils usually is sufficient to retain the applied ammonium near the surface. Excessive hydraulic loading in rapid infiltration systems will eventually saturate the ammonium adsorption capacity and permit downward movement of ammonium. Retention of ammonium in the exchangeable form is temporary in any case because the adsorbed ammonium is nitrified when oxygen becomes available in the drying cycle. The exchangeable ammonium, however, plays a very important role in the nitrification-denitrification sequence by holding nitrogen near the soil surface until the environment becomes aerobic during drying.

Nitrification. As noted before, most of the nitrogen in treated wastewater is in the form of ammonium ion. Consequently, when wastewater containing ammonium is discharged to the environment, depletion of oxygen resources in receiving water or in soils can occur as the ammonium is oxidized to nitrite and nitrate. The conversion of ammonium ion to nitrite and nitrate in soils and receiving waters is primarily achieved by certain autotrophic bacteria, particularly nitrifying bacteria, which oxidize noncarbonaceous matter for energy. The nitrifying bacteria usually are present in relatively small numbers in untreated domestic wastewater, but these bacteria, particularly *Nitrosomonas* and *Nitrobacter* are common soil inhabitants and are usually present in sufficient numbers to convert added ammonia to nitrate rapidly and completely. The principal stoichiometric reactions involved in the conversion of ammonium ion to nitrite and nitrate are shown in the following equations (Equations 3 and 4):

$$2NH_4^+ + 3O_2 \xrightarrow{\text{Nitrosomonas}} 2NO_2^- + 2H_2O + 4H^+ \quad (3)$$

and

$$2NO_2^- + O_2 \xrightarrow{\text{Nitrobacter}} 2NO_3^- \quad (4)$$

Along with these energy reactions, some of the ammonium ions are assimilated into cell tissue and the overall reaction proposed to describe the autotrophic conversion of ammonium ion to nitrate is as follows:[2]

$$22NH_4^+ + 37O_2 + 4CO_2 + HCO_3^- \rightarrow$$
$$C_5H_7NO_2 + 21NO_3^- + 20H_2O + 42H^+ \quad (5)$$

Although nitrifying bacteria are present in sufficient numbers in most soils, microbial populations may be initially low in subsoils or in coarse-textured soils that are prone to be dry much of the time. In such soils, several weeks may be required for nitrifying bacteria to attain a population needed for nitrification.

Rates of nitrification are affected by soil temperature, pH, dissolved oxygen concentration, microbial population, and ammonium ion concentrations in applied wastewater. Nitrification kinetics are, however, much more likely to be controlled by lack of oxygen and low temperature in soils than by insufficient populations of nitrifying bacteria. The usual situation in soils is that nitrite rarely accumulates, suggesting that the rate-controlling step is in the oxidation of ammonium ion to nitrite. Nitrite oxidation is inhibited by molecular ammonia in liquid systems, particularly when the solution pH is alkaline. However, in soil systems, adsorption of ammonium near the soil surface tends to prevent this inhibitory effect from becoming a practical difficulty in most circumstances.[3]

Under favorable soil moisture and temperature conditions, measured values of ammonium ion oxidized to nitrate, ranging daily from 5 to 50 mg/L N, have been reported. Assuming that the nitrification process occurs in the top 10 cm (4 in.) of soils because of ammonium ion adsorption, approximately 7 to 67 kg/ha·d (6 to 60 lb/d/ac) of ammonium nitrogen is converted to nitrate.

These figures indicate that nitrification will occur even with application rates (of wastewater containing 20 mg/L NH_4-N) in the range of 3 to 30 cm/d (1.2

to 12 in./d). Higher nitrification rates in soil columns also have been reported.[3]

These calculations are consistent with field observations—that complete oxidation of ammonium ion contained in applied wastewater occurs if wastewater application periods are short enough to prevent soils from becoming anaerobic.

The theoretical oxygen requirement for nitrification is about 4.6 mg oxygen/mg of ammonium-N oxidized. Although the nitrifying bacteria are obligate aerobes, they will continue to function at oxygen concentrations well below the minimum dissolved oxygen concentration required in wastewater treatment systems. The optimum soil temperature for nitrification is in the range of 24° to 35°C (75° to 95°F). However, some nitrification has been reported at soil temperatures as low as 2°C (36°F). The optimum pH for nitrification is in the neutral-to-slightly alkaline range and the controlling factor is the pH of the soil. Nitrification falls off sharply in acid soils with a limiting value in the neighborhood of pH 4.5.

NITROGEN REMOVAL

All land treatment processes affect nitrogen concentrations, with slow rate (irrigation) and overland flow the more efficient processes in nitrogen removal. These two processes can remove 75 to 90% of applied nitrogen. Rapid infiltration is less efficient, averaging 50% typically; with special management, however, it can also remove 65 to 90% of the applied nitrogen.

Removal mechanisms. There are three removal mechanisms for nitrogen in soil systems—denitrification, crop uptake, and volatilization. For slow rate systems, crop uptake and denitrification are most important, although ammonia volatilization can be significant. For rapid infiltration, denitrification is the only important mechanism. For overland flow, both denitrification and crop uptake can be important.

Denitrification. Denitrification is a biological process of converting nitrate nitrogen into nitrogen gas. Denitrifying bacteria are facultative organisms that prefer oxygen but can use nitrate and nitrite as electron acceptors under oxygen-deficient conditions. Denitrifying bacteria are common soil organisms that require oxygen-deficient or anaerobic microsites and the presence of organic matter to function.[4] It has been shown that denitrification can occur in soils that are considered to be aerobic.[3] This occurs because temporary saturation of soil pores with water or wastewater will cause rapid depletion of oxygen. These oxygen-deficient microsites within soils are common features of virtually all soils.

The denitrification reaction proceeds slowly at temperatures of 2°C (36°F). Denitrification is also slow in acid soils, increasing rapidly when soil pH is neutral to slightly alkaline. The process requires 3.2 g of carbon source (expressed as glucose) for each gram of nitrogen denitrified. It is therefore important to apply a wastewater with a BOD to nitrogen ratio of 3 or greater to drive the denitrification reaction.

Crop uptake. Crops vary in their efficiency in removing nitrogen from the soil-water solution. Typically, crops will remove 35 to 60% of the available nitrogen in the soil-water.[3] This depends on the crop, the depth and distribution of rooting, nitrogen loading rate, rate of water movement through the soil, and other factors. A plot of nitrogen uptake versus nitrogen loading is presented in Figure 9.1 for coastal bermuda grass grown with wastewater in Tallahassee, Fla.[5]

Grasses are more efficient in nitrogen uptake than row crops. Total quantities of nitrogen taken up by crops generally fall within the range of 50 to 500 kg/ha·a (45 to 450 lb/yr/ac). To be ef-

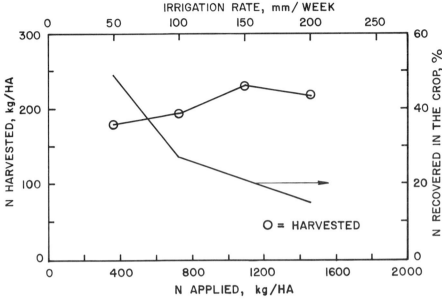

FIGURE 9.1. Nitrogen harvested by coastal bermuda grass.

fective, the crop containing nitrogen must be harvested.

Volatilization. Ammonia nitrogen can be lost to the atmosphere by volatilization during sprinkler application or from the soil surface during the drying period. Volatilization requires contact with air and conversion of ammonium ions to free ammonia. The equilibrium between ammonium ion and ammonia is regulated by pH, and the proportion of free ammonia is small at neutral pH values. At Hanover, N.H., with a wastewater of pH 7.3, the ammonia volatilization was measured at 3 to 5%.

Nitrogen removal by slow rate systems. Nitrogen removal in slow rate systems results from a combination of crop uptake, denitrification, soil storage, and volatilization. Crop uptake is the primary removal mechanism, but denitrification can be significant. Different removals are expected for agricultural irrigation, where the crop nitrogen can be physically removed through harvesting, as compared with landscape and woodlands irrigation.

Agricultural irrigation. If total nitrogen input does not greatly exceed crop requirements for nitrogen, removals of 35 to 60% can be expected as a result of crop uptake, depending on the crop. Depending on soil properties and irrigation schedules, denitrification may account for 15 to 70% of the applied nitrogen. In nonwastewater agricultural practice where attempts are made to minimize denitrification, losses of 15 to 30% are common.[3]

Removals of nitrogen in slow rate systems are presented in Table 9.1. Except for the Hanover system, the systems represent full-scale operations with corn, grass, and other forages being grown. At Hanover, the demonstration test cells were dosed with wastewater throughout the year despite the fact that the climate would normally dictate 3 to 5 months storage for forage crop irrigation. The applied nitrogen was 590 to 610 kg/ha·a (525 to 540 lb/yr/ac), which is significantly more than the forage grasses could remove. The low nitrogen removal at San Angelo can be explained by the fact that commercial fertilizer is

TABLE 9.1. Nitrogen removal in slow rate systems.

Location	Total nitrogen applied kg/ha·a	mg/L	Percolate total nitrogen mg/L	Removal, %
Dickinson, N.D.	150	11.8	3.9	67
Hanover, N.H.				
Primary effluent	610	28.0	9.5	66
Secondary effluent	590	26.9	7.3	73
Melbourne, Australia	880	50	5.9	88
Muskegon, Mich.	130	8.2	2.5	70
Pleasanton, Calif.	630	27.6	2.5	91
Roswell, N.M.	465	66.2	10.7	84
San Angelo, Tex.	830	35.4	6.1	83

applied in addition to the 830 kg/ha·a (740 lb/yr/ac) of wastewater nitrogen.

The concentration of nitrogen in the percolate at Hanover was correlated with nitrogen loading, as presented in Figure 9.2. If the percolate nitrogen is to be maintained at or below 10 mg/L, the Hanover system, where the crop is a mixture of reed canarygrass, timothy, smooth bromegrass and later, orchardgrass, should not be loaded at more than 800 kg/ha·a (712 lb/yr/ac).[6]

Landscape irrigation. There are many cases of golf course, park, and open space irrigation in which vegetation may be harvested, but is not physically removed. The effective long-term nitrogen removal of these systems is unknown.

Woodlands irrigation. Nitrogen removal in woodlands can be significant, through a combination of denitrification and crop uptake. Young, developing forests have the maximum ability to use nitrogen. As they become mature, forests recycle their nitrogen in decaying organic matter, which requires denitrification or soil accumulation for effective nitrogen removal. Several wastewater applications to woodlands are summarized in Table 9.2.[7] Nitrogen removal depends on the species irrigated and whether young trees are grown and harvested regularly. Young plantations

of Douglas fir and poplar can take up 250 kg/ha·a (223 lb/yr/ac).

Nitrogen removal by rapid infiltration systems. Rapid infiltration systems are traditionally the least effective of the land treatment systems for nitrogen removal. As denitrification is the key to nitrogen removal, the important design criteria are BOD:N ratio, loading rate, and ratio of flooding period to drying period. The design objective is to man-

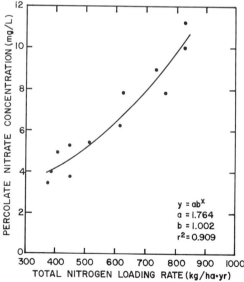

FIGURE 9.2. Percolate nitrate-nitrogen concentration from Hanover test cells.

TABLE 9.2. Nitrogen loading and removal in forests.

| Forest type | Tree age, yr | Total nitrogen, kg/ha·a | | | Nitrate nitrogen in leachate, mg/L |
		Loading	Tree uptake	Grass uptake	
Douglas Fir seedling	4	350	111	121	5.3
Poplar (Northwest)	4	400	166	173	0.1
Poplar (Lake States)	5	103	68	—	2.8
Red Pine (Lake States)	20–25	131	—	—	5.4
Southern mixed hardwood	45	684	—	—	8.0[a]
Eastern mixed hardwood	70	150	95	—	4.6

[a] At the base of the slope where significant denitrification occurs, the concentration is 3.7 mg/L.

age one or more of these factors to achieve sequential or simultaneous nitrification/denitrification to remove the nitrogen as a gas. Manipulating the BOD:N ratio and flooding/drying period will provide the needed conditions for denitrification. The loading rate, if kept to within the range of 15 to 30 m/a (50 to 100 ft/yr), will provide adequate detention time for effective nitrogen removal. Nitrogen removal experience in rapid infiltration is summarized in Table 9.3.[8]

At Brookings, S.D., the low infiltration rate of clay loam soil limits the liquid loading and provides adequate detention time in the soil profile. At Boulder, Colo., the high infiltration rate and the shallow underdrains do not allow adequate detention time for denitrification. The pilot system is designed for nitrification and therefore no attempt is made to provide nitrogen removal.

The highest nitrogen removals are associated with the lowest hydraulic loading rates and the highest BOD:N ratios. This indicates that intermediate rates (15 to 30 m/a) are more conducive to nitrogen removal.

Nitrogen removal by overland flow systems. The overland flow system is ideally suited to the nitrification-denitrification sequence, which requires aerobic and anaerobic zones in close proximity. A double layer is formed on the soil surface with the upper layer aerobic and the low layer oxygen-deficient. Applied wastewater can be nitrified readily because of the contact with atmospheric oxygen as the wastewater flows over the soil surface. Nitrates thus formed diffuse into the oxygen-deficient layer, encountering reducing conditions in which denitrification can proceed. Denitrification is enhanced by a BOD to nitrogen ratio of 3 or more.

TABLE 9.3. Nitrogen removal at rapid infiltration systems.

| | Total nitrogen applied | | Percolate nitrogen mg/L | BOD:N ratio | Removal, % |
	kg/ha·a	mg/L			
Boulder, Colo.	8 050	16.5	9–16	2.3:1	0–20
Brookings, S.D.	1 330	10.9	6.2	2:1	43
Calumet, Mich.	4 170	24.4	7.1	3.4:1	71
Fort Devens, Mass.	15 250	50	10–20	2.4:1	60–80
Hollister, Calif.	6 110	40.2	2.8	5.5:1	93
Lake George, N.Y.	6 960	12	7.5	2:1	38
Phoenix, Ariz.	16 710	27.4	9.6	1:1	65

Nitrogen removals from three overland flow projects are summarized in Table 9.4. At Ada, Okla., the climate allowed year-round application. The summertime removals of nitrogen were greater than wintertime removals by the amount credited to crop nitrogen. At Hanover, the reported removals are for May through October, although applications continued through the winter. Nitrogen removal in winter dropped off to about 25%. Storage would normally be used in this northern climate.

BOD REMOVAL

Land treatment systems can remove organics from wastewater at rates comparable to aerated lagoon systems. Aerated lagoons can remove BOD at rates of 600 to 2 000 kg/ha·d (534 to 1 780 lb/d/ac) or more and land treatment systems have similar capabilities. It has been shown that soils containing bacteria acclimated to organic wastewaters can accept 11 200 kg/ha·d (10 000 lb/d/ac) of COD without becoming overloaded.[9] Loading rates in this range in actual applications can lead to odors if systems are not managed carefully. Removals for the three land treatment systems are discussed separately below. Because organic loading rates above 200 kg/ha·d (178 lb/d/ac) are uncommon with domestic wastewater, some experience with land treatment of food processing wastewaters where higher rates are common will be discussed.

Acceptable organic loading rates depend on the ability of the soil to drain and on the drying period. For slow rate and rapid infiltration systems, the drainage of water is vertical through the soil profile, and the drying period is typically 6 to 20 days. For overland flow systems, the drainage is provided by the surface slope and drying periods are typically 16 to 20 hours.

Food processing wastewater. For slow rate systems applying screened, raw cannery wastewater, organic loading rates are typically 50 to 500 kg/ha·d (45 to 446 lb/d/ac) of BOD.[10] The limiting parameter is usually the hydraulic loading rate for the specific soil, climate, and crop water requirement. A monitoring program was conducted for slow rate land treatment in a study of five potato processing wastewater systems in Idaho.[11] Percolate water quality at various depths was measured. It was

TABLE 9.4. Nitrogen removal in overland flow systems.

Parameter	Location		
	Ada, Okla.	Hanover, N.H.	Utica, Miss.
Type of wastewater	Raw screened	Primary	Pond effluent
Hydraulic loading, cm/wk	10	5	6
BOD to N ratio	6.3	2.3	1.1
Nitrogen, kg/ha·a			
Applied	1 200	950	660
Removed	1 100	890	500–600
Crop uptake	100	210	250
Denitrification[a]	1 000	680	250–350
Removal, mass basis, %	91	94	75–90
Total nitrogen, mg/L			
Applied	23.6	36.6	20
Runoff	2.2	5.4	2–5
Removal, concentration basis, %	91	85	75–90

[a] Denitrification is the primary mechanism responsible for unmeasured losses, but this value includes volatilization, soil storage, and percolation.

found that the mean COD removal ranged from 95 to 98% at the 150-cm (5-ft) depth. The loading rates ranged from 45 to 315 kg/ha·d (40 to 280 lb/d/ac).

The typical BOD loading rates for rapid infiltration systems vary from 150 to 1 000 kg/ha·d (134 to 890 lb/d/ac). The EPA recommended limit is 674 kg/ha·d (600 lb/d/ac) for these systems.[12]

Overland flow systems typically operate at BOD loadings of 40 to 110 kg/ha·d (36 to 98 lb/d/ac). Removals of 98% or more can be expected at these rates. With high strength wastewater, it has been found that recycling of effluent to reduce applied BOD concentrations to below 1 000 mg/L was effective. BOD loading rates were 56 to 112 kg/ha·d (50 to 100 lb/d/ac).[13]

Municipal wastewater. Relatively few municipal land treatment systems (except for overland flow) have been monitored for BOD concentrations in the treated water. The following data, therefore, represent both full-scale and demonstration systems.

Slow rate systems. On the basis of the previous discussion of industrial slow rate systems operating with applied BOD concentrations of 1 000 mg/L or more, it is easy to see that municipal wastewater (at 200 mg/L BOD) or primary effluent will not create problems for carbon removal. Instead of 50 kg/ha·d (45 lb/d/ac), the BOD loading rates for municipal systems using secondary effluent are typically less than 6 kg/ha·d

(5 lb/d/ac). Four full-scale municipal slow rate systems and a demonstration system at Hanover, N.H., are presented in Table 9.5. The removal rates are in terms of concentration reduction, not mass removal. The percolate concentrations support the expected value of less than 2 mg/L, as presented in the U.S. EPA Process Design Manual.[8]

Rapid infiltration systems. The percolate water quality from rapid infiltration systems is nearly as high as from slow rate systems for BOD despite the fact that 10 times more organic material is typically applied. As shown in Table 9.6, the percolate BOD is typically near 2 mg/L.[8,14] At Hollister, Calif., the two shallow wells monitoring the percolate average 6 and 13 mg/L BOD. The systems at Brookings, S.D., and at Phoenix, Ariz., are demonstration projects.

Overland flow systems. Overland flow systems can also operate at higher organic loadings than the 5 to 50 kg/ha·d rates that are typical with municipal wastewater.[10,13] Removals of BOD for three research and demonstration projects are presented in Table 9.7.

PHOSPHORUS REMOVAL

Biological, chemical, or physical immobilization of phosphorus and plant uptake are the main mechanisms for phosphorus removal. The usual forms of phosphorus found in municipal wastewater include orthophosphate, polyphos-

TABLE 9.5.   BOD removal in slow-rate systems.

Location	Applied wastewater, kg/ha·d	mg/L	Percolate, concentration mg/L	Removal, %
Dickinson, N.D.	3	42	<1	>98
Hanover, N.H.				
Primary	11	101	1.4	98.6
Secondary	4	36	1.2	96.7
Muskegon, Mich.	3	24	1.3	94.6
Roswell, N.M.	3	43	<1	>98
San Angelo, Tex.	10	119	1.0	99.1

TABLE 9.6.   BOD removal for rapid infiltration systems.

Location	Applied wastewater BOD kg/ha·d[a]	Applied wastewater BOD mg/L	Percolate concentration, mg/L	Removal, %
Boulder, Colo.	54[b]	131[b]	10[b]	92
Brookings, S.D.	13	23	1.3	94
Calumet, Mich.	107[b]	228[b]	58[b]	75
Fort Devens, Mass.	87	112	12	89
Hollister, Calif.	177	220	8	96
Lake George, N.Y.	53	38	1.2	97
Milton, Wis.	155	28	5.2	81
Phoenix, Ariz.	45	15	0–1	93–100
Vineland, N.J.	48	154	6.5	96

[a] Total kg/ha·a applied divided by number of days in the operating season.
[b] COD basis (14)

phate, and organic phosphate. The various forms of orthophosphates, PO_4^{3+}, HPO_4^{2-}, $H_2PO_4^-$, and H_3PO_4, are available for biological metabolism without further breakdown. Polyphosphates include those molecules with two or more phosphorus, oxygen, and in some instances, hydrogen atoms combined in a complex molecule. Polyphosphates undergo hydrolysis in aqueous solutions and convert to the orthophosphate forms. However, the hydrolysis reactions may be quite slow in the soil environment. Organic phosphate is usually of minor importance in most municipal wastewater, but it can be an important constituent of industrial wastes and wastewater sludge.

Typical data on the phosphorus concentrations in untreated municipal wastewater are in the range of 4 to 15 mg/L as total P, with 3 to 10 mg/L P as inorganic phosphorus and 1 to 5 mg/L P as organic phosphorus. With most wastewater, approximately 10% of the phosphorus, corresponding to the portion that is insoluble, is removed by primary treatment.[15] Except for the amount incorporated into biomass, the additional removal achieved in conventional biological treatment is minimal because almost all the phosphorus present after primary sedimentation is in a soluble form.

In land treatment, phosphorus as orthophosphate reacts with practically all soils with an almost quantitative removal from solution.[1] Because soils characteristically have very reactive surfaces containing iron, aluminum, and calcium, insoluble phosphates are formed with these species in the soil environment. Acidic conditions favor Fe-P and Al-P complexes, and alkaline conditions

TABLE 9.7.   BOD removal in overland flow demonstration systems.

Location	Applied wastewater kg/ha·d	Applied wastewater mg/L	Runoff concentration mg/L	Removal, %
Ada, Okla.	24	150	8	94.7
Hanover, N.H.				
Primary	20	72	9	87.5
Secondary	5	45	5	88.8
Utica, Miss.	11	22	8	63.6

Note: kg/ha·d × 0.89 = lb/ac·d

TABLE 9.8. Annual uptake of elements for forests and crops in relation to quantities applied with wastewater.

Element	Wastewater, kg/ha[a]	Forest kg/ha	Forest %[b]	Crops kg/ha	Crops %[b]
N	395	22–90	14	45–336	48
P	197	2–22	6	2–50	13
K	276	3–73	14	56–336	71
Ca	472	7–56	7	11–168	19
Mg	334	3–9	2	2–56	9
Na	983	1–3	0.2	2–45	2
Fe	10	0.1–0.06	3	<1	11
Mn	10	0.6–1	8	<1	11
Cu	5	<0.06	<1	<0.1	<3
Zn	6	<0.01	<2	<0.1	<3

[a] These figures are based on an average composition of effluent and on the assumption that 198 cm (78 in.) of effluent is applied per year.
[b] Percent of total applied.

favor Ca-P retention.[16] Retention of organic phosphorus at colloidal surfaces may also take place and alter their chemical and biological stability. The retention of phosphorus by aluminum and iron apparently involves both precipitation (reaction with iron and aluminum cations in solution) and adsorption (reaction at the surfaces of iron and aluminum compounds) where adsorption predominates at low phosphorus concentrations. Mechanisms of phosphorus retention by calcium are similar, involving the formation of more insoluble compounds, such as $Ca_4OH(PO_4)_3 \cdot H_2O$ and $Ca_{10}F_2(PO_4)_6$ or adsorption and occlusion of phosphorus onto calcium carbonate precipitates.[16,17]

The crop itself can remove certain amounts of phosphorus and other macro- and micro-nutrients and incorporate them in the biomass, as shown in Table 9.8.[17] In normal agricultural practices, the phosphorus recovery from fertilizers in crops is about 30%.

The efficiency of phosphorus removal obtained with slow rate land treatment systems on a variety of soils is summarized in Table 9.9.[8] In general, removal is primarily affected by soil type and depth of soil column, design and mode of application of wastewater, and vegetative cover. The phosphorus removal

increases with the clay content of the soil or decrease of the permeability. Phosphorus is effectively removed in the upper 30 to 60 cm (1 to 2 ft) of the soil column by adsorption/precipitation reactions when clay, oxides of iron and aluminum, and calcareous materials are present. It is estimated that every 10 years a depth of 30 cm (1 ft) will be saturated with phosphorus, which in the long run can limit the use of the soil.

Phosphorus removal data for rapid infiltration systems are presented in Table 9.10.[8] Phosphorus removal in rapid infiltration depends on soil type and pH, travel distance, and wastewater loading rates.[8]

For overland flow, phosphorus removal is limited by the lack of complete contact between the wastewater and the adsorption sites within the soil. As a result of this limited soil contact, phosphorus removals achieved at existing overland flow systems generally range from 40 to 60%. Phosphorus removal data for overland flow systems are shown in Table 9.11.[8]

Phosphorus removal in overland flow can be enhanced by chemical addition prior to application. At Ada, Okla., addition of alum at an Al:P mole ratio of 2:1 was shown to reduce total phosphorus in the runoff to 1.0 mg/L.[18] At Utica,

189

TABLE 9.9. Phosphorus removal data for typical slow rate systems.

Agricultural system locations	Annual wastewater loading rate, cm/yr	Surface soil	PO₄ concentration in applied wastewater, mg/L as P	Soluble PO₄ concentration in affected groundwater, mg/L as P	Removal, %	Sampling depth, m	Distance from application site, m	Soluble PO₄ concentration in background groundwater, mg/L as P
Camarillo, Calif.	160	Clay loams and sandy loams	11.8[a] 11.8[a]	2.8[a] 0.2[a]	76[a] 98[a]	1 3	0 0	3.0[a] —
Dickinson, N.D.	140	Sandy loams and loamy sands	6.9[a]	0.05[a]	99[a]	<5	30–150	0.04[a]
Hanover, N.H.	130–78	Sandy loam and silt loam	7.3–7.6[a]	0.03–0.07[b]	99.0–99.5	1.5	0	—
Mesa, Ariz.	400–860	Loamy sands and sandy loams	9.0[b] 9.0[b]	5.0[b] 4.2[b]	44[b] 53[b]	0.5 1	0 0	1.0[b] 3.6[b]
Muskegon, Mich.	130–260	Sands and loamy sands	1.0–1.3[a]	0.03–0.05[a]	95–98[a]	1.5	0	0.03[a]
Roswell, N.M.	80	Silty clay loams	7.95[a]	0.39[a]	95[a]	<6	0	0.55[a]
Tallahassee, Fla.		Sand						
Winter	520		10.5[a]	0.1[a]	>99[a]	1.2	0	0.02[a]
Summer	1040		10.5[a]	0.0[a]	>99[a]	10.7	0	0.02[a]
Forest systems locations								
Helen, Ga.	380	Sandy loam	13.1[a]	0.22[a]	98[a]	1.2	0	0.21[a]
State College, Pa. (Penn State University)	260	Sandy loams and clay loams	7.7[b]	0.08[b]	99[b]	1.2	0	0.03[b]

[a] Total phosphate concentration.
[b] Orthophosphate concentration.

TABLE 9.10. Phosphorus removal data for selected rapid infiltration systems.[8]

Location	Average concentration in applied wastewater, mg/L	Distance of travel, m		Average concentration in renovated wastewater, mg/L	Removal, %
		Vertical	Horizontal		
Boulder, Colo.[a]	6.2	2.4-3.0	0	0.2-4.5	40-97
Brookings, S.D.[b]	3.0	0.8	0	0.45	85
Calumet, Mich.[a]	3.5	3-9	0-125	0.1-0.4	89-97
	3.5	—[c]	1700[c]	0.03	99
Fort Devens, Mass.[b]	9.0	15	30	0.1	99
Hollister, Calif.[b]	10.5	6.8	0	7.4	29
Lake George, N.Y.[b]	2.1	3	0	<1	>52
	2.1	—[c]	600[c]	0.014	99
Phoenix, Ariz.[a]	8-11	9.1	0	2-5	40-80
	7.9	6	30	0.51	94
Vineland, N.J.[b]	4.8	2-18	0	1.54	68
	4.8	4-16	260-530	0.27	94

[a] Total phosphate measured.
[b] Soluble phosphate measured.
[c] Seepage.

Miss., mass removals ranged between 65 and 90% with alum addition as compared to less than 50% without alum. Effluent concentrations ranged from 1.3 to 4 mg/L.[19]

MICRONUTRIENT REMOVAL

Micronutrients such as zinc, iron, manganese, and copper are removed by the soil through the mechanisms of surface sorption, ion exchange, precipitation, and surface complex ion formation. As with phosphorus removal, the mechanisms are time-dependent and it is not possible to predict the actual renovative capacity of a site based on simple ion exchange or soil adsorption theories.

Slow rate systems. Slow rate systems are efficient in micronutrient and metal removal. Depending on the concentration in the applied wastewater, removals can be as high as 95%. Reported removals of selected micronutrients and metals for Muskegon, Mich., San Angelo, Tex., and Melbourne, Australia are presented in Table 9.12.[20]

Rapid infiltration systems. Removal of trace metals by rapid infiltration has been monitored at several sites with long histories of operation. A comparison of concentrations in the applied wastewater and in the groundwater is presented in Table 9.13.[21–23] At Hollister, Calif., after 33 years of application, the

TABLE 9.11. Phosphorus removal at overland flow systems.[8]

Location	Hydraulic loading rate, cm/d	Total phosphorus, mg/L	
		Applied	Treated runoff
Ada, Okla.	3.3	8.0	4.5
Easley, S.C.	2.36	8.9	4.0
Hanover, N.H.	2.8	6.6	4.4
Utica, Miss.	2.54	10.3	5.9

TABLE 9.12. Micronutrient and heavy metal removal by slow rate systems.[20]

Constituents	Muskegon, Mich. Percolate concentration, mg/L	Removal, %	San Angelo, Tex. Percolate concentration, mg/L	Removal, %	Melbourne, Australia Percolate concentration, mg/L	Removal, %
Cadmium	<0.002	90	<0.004	—[a]	0.002	80
Chromium	0.004	90	<0.005	>98	0.03	90
Copper	0.002	90	0.014	85	0.02	95
Lead	<0.050	>40	<0.005	—[a]	0.01	95
Manganese	0.26	15	—	—	—	—
Mercury	<0.002	—[a]	—	—	0.0004	85
Zinc	0.033	95	0.102	25	0.04	95

[a] Not calculated because influent and percolate values are below detection limits.

shallow groundwater beneath the site showed limited removal of copper and slight increases in concentrations of zinc, cobalt, and iron. Manganese was being leached out of the soil profile and was found in an average concentration of 0.96 mg/L in the groundwater. This exceeds the irrigation water quality guidelines of 0.2 mg/L.

At Phoenix, Ariz., the rapid infiltration system removed substantial quantities of copper and zinc but not of cadmium or lead. In general, rapid infiltration systems do not achieve high removals of micronutrients.

Overland flow systems. Heavy metal and micronutrient removal for overland flow should be similar to phosphorus removal, although little information exists. One study with low application rates showed removals of 90 to 98%, which is higher than would be expected based on a comparison with phosphorus removal data.

GROUNDWATER PROTECTION AND MONITORING

Monitoring of land treatment systems involves the observation of significant changes resulting from the application of wastewater. The monitoring data are used to confirm environmental predictions and to determine if any corrective action is necessary to protect the environment or maintain the renovating capacity of the system. The components of the environment that need to be observed include wastewater, groundwater, and soils on which wastewater is applied and, in some cases, vegetation growing in soils receiving wastewater.[3]

TABLE 9.13. Trace element concentrations at long-term rapid infiltration systems, mg/L.[21-23]

Constituent	Calumet, Mich. Applied	Groundwater	Hollister, Calif. Applied	Groundwater	Milton, Wis. Applied	Groundwater
Cadmium	0.026	0.022	<0.004	0.028	0.020	0.020
Chromium	0.027	0.021	<0.014	<0.017	0.050	0.050
Copper	0.043	0.022	0.034	0.038	0.050	0.050
Lead	0.005	0.005	0.054	0.090	0.100	0.100
Manganese	—	—	0.070	0.960	0.200	0.200
Nickel	0.005	0.006	0.051	0.130	0.100	0.290
Zinc	—	—	0.048	0.082	0.078	0.254

Water quality management. Monitoring of water quality for land application systems is generally more involved than for conventional treatment systems because nonpoint discharges of system effluent into the environment are involved. Monitoring of water quality at several stages of a land treatment process may be needed for process control. These stages may be: applied wastewater, renovated water, and receiving waters—surface water or groundwater.

Applied wastewater. The water quality parameters and the frequency of analyses will vary from site to site depending on the regulatory agencies involved and the nature of the applied wastewater. The measured parameters may include: those that may adversely affect receiving water quality either as a drinking water supply or an irrigation water supply; those required by regulatory agencies; and those necessary for system control. An example of a possible water quality monitoring program for a slow rate system is presented in Table 9.14.

Applied wastewater, soil, and groundwater should be tested initially and periodically thereafter, as appropriate, for trace elements or other constituents of environmental concern. In the arid West the total dissolved solids and sodium adsorption ratio (SAR) are important. But these may not be important in areas with humid climate. For small systems, the number of groundwater wells will usually be limited. In addition, monitoring of constituents such as suspended solids and phosphorus may be eliminated for some systems.

Renovated water. Renovated water may be recovered as runoff in an overland flow system, or as drainage from underdrains or groundwater from recovery wells in slow rate and rapid infiltration systems. Point discharge to surface waters must satisfy National Pollutant Discharge Elimination System (NPDES) permit.

Groundwater. In groundwater, travel time of constituents is slow and mixing is not significant compared with surface waters. Surface inputs near a sampling well will move vertically and arrive at the well much sooner than inputs several hundred feet away from the well. Thus, the groundwater sample represents contributions from all parts of

TABLE 9.14. Example monitoring program for a slow-rate system.

	Frequency of analysis				
			Groundwater		
Parameter	Applied wastewater	Soil	Onsite wells	Perimeter wells	Background wells
Flow	C	—	—	—	—
BOD or TOC	M	—	Q	Q	Q
Suspended solids	M	—	—	—	—
Nitrogen, total	W	A	Q	Q	Q
Nitrogen, nitrate	—	—	Q	Q	Q
Phosphorus, total	Q	A	—	—	—
Coliforms, total	W	—	Q	Q	Q
pH	W	A	Q	Q	Q
Total dissolved solids	M	—	Q	Q	Q
Alkalinity	M	—	Q	Q	Q
SAR (sodium adsorption ratio)	M	A[a]	Q	Q	Q
Static water level	—	—	M	M	M

Note: C = Continuously; Q = Quarterly; A = Annually; W = Weekly; and M = Monthly.
[a] Exchangeable sodium percentage measured in the soil.

the surface area with each contribution arriving at the well at a different time. A sample may reflect surface inputs from several years before sampling and have no association with the land application system. Consequently, it is imperative to obtain adequate background water data and to locate sampling wells so that response times are minimized.

If possible, existing background data should be obtained from wells in the same aquifer both beyond and within the anticipated area of influence of the land application system. Wells with the longest history of data are preferable. Monitoring of background wells should continue after the system is in operation to provide a base for comparison.

In addition to background sampling, samples should be taken from groundwater at perimeter points in the downgradient direction of groundwater movement from the site. In locating the sampling wells, consideration must be given to the position of the groundwater flow lines resulting from the application. Perimeter wells should be located sufficiently deep to intersect flow lines emanated from below the application area, but not so deep as to prolong response times.

In addition to quality, the depth to groundwater should be measured at the sampling wells to determine if the hydraulic response of the aquifer is consistent with what was anticipated. For slow rate systems, a rise in water table levels to the root zone would necessitate corrective action such as reduced hydraulic loading or added underdrainage. The appearance of seeps or perched groundwater tables might also indicate the need for corrective action.

Soils management. In some cases, application of wastewater to the land will result in changes in soil properties. Results of soil sampling and testing will serve as the basis for deciding whether or not soil properties should be adjusted by the application of chemical amend-

ments. Soil properties that are important to management include: pH, salinity, nutrients and trace elements, and exchangeable cations.

pH. Soil pH below 5.5 or above 8.5 generally is harmful to most plants. Below pH 6.5, the capacity of soils to retain metals is reduced significantly. A soil pH above 8.5 generally indicates a high sodium content and possible permeability problems. If wastewaters contain high concentrations of sodium, the soil pH may rise in the long term. A pH adjustment program should be based on the recommendations of a professional agricultural consultant or county or state farm advisor.

Salinity. The levels at which salinity becomes harmful to plant growth depend on the type of crop. Salinity in the root zone is controlled by leaching soluble salts to the subsoil or drainage system.

Nutrient and trace element. The nutrient status of the soil and the need for supplemental fertilizers should be periodically assessed. The levels of metals in the soil may be the factor determining the ultimate useful life of the system. University agricultural extension services may provide the service or recommend competent laboratories.

Exchangeable cations. High levels of sodium in the soil cause low soil permeability, poor soil aeration, and difficulty in seedling emergence.[8] It is important to maintain high levels of exchangeable calcium and magnesium in soil and low levels (5% or so) of exchangeable sodium. Fine textured soils may be affected at an exchangeable sodium percentage (ESP) of 10%, but coarse-textured soil may not be damaged until the ESP reaches about 20%.

Vegetation management. Plant tissue analysis is probably more revealing

TABLE 9.15. Comparison of annual costs of slow-rate land treatment versus conventional alternatives, cents/m³.

Costs[a]	Slow rate	Secondary	AWT with phosphorus and nitrogen removal
Capital	54	25	54
Operation and maintenance	9	29	48
Total	63	54	102
Local share	17[b]	35[c]	62[c]

[a] For a flow of 37 850 m³/d (10 Mgal/d), cents/m³ × 3.785 = cents/1000 gal. Costs are March 1978 dollars (ENR 2693).

[b] 15% of capital plus 100% of operation and maintenance.

[c] 25% of capital plus 100% of operation and maintenance.

than soil analysis with regard to deficient or toxic levels of elements. All of the environmental factors that affect the uptake of an element are integrated by the plant, thus eliminating much of the complexity associated with interpretation of soil test results. If a regular plant tissue monitoring program is established, deficiencies and toxicities can be determined and corrective action can be taken.

COSTS OF LAND TREATMENT

The capital cost of a land treatment system depends on the field area required and the cost of the land as well as the cost of preapplication treatment, transmission, and storage. Because slow rate systems require the most land, they generally have the highest capital cost. The overall costs (capital plus operating) of land treatment systems are generally comparable or less than those of conventional secondary or advanced wastewater treatment.[24] A comparison of costs of slow rate land treatment to secondary and advanced wastewater treatment is shown in Table 9.15.[25]

Operation and maintenance (O&M) costs include power, labor, and materials and can be partially offset by revenues such as from the sale of crops or lease of land. Graphs and equations of capital and O&M costs for flow rates between 379 m³/d and 378 500 m³/d (0.1 to 100 Mgal/d) are available in "Cost of Land Treatment Systems."[26]

To allow typical costs to be compared for slow rate, rapid infiltration, and overland flow, a series of assumptions have been made and are presented in Table 9.16.[24] Costs for 1893 m³/d and 189 250 m³/d (0.5 and 50 Mgal/d) systems are compared in Table 9.17.[24] These costs are for April 1980 (ENR 3143). As indicated in Table 9.17, the slow rate system is the most expensive of the three systems. The O&M costs of slow rate are approximately 20% of the total costs. This is because total O&M costs are partially offset by the revenues from crop production.

TABLE 9.16. Land treatment system assumptions for cost components.[24]

Component	Slow rate	Rapid infiltration	Overland flow
Preapplication treatment	Oxidation ponds	Oxidation ponds	Oxidation ponds
Transmission length, km			
1893 m³/d	0.6	0.6	0.6
189 250 m³/d	2.6	2.6	2.6
Storage	90 days	7 days	35 days
Distribution	Surface flooding	Basins	Sprinklers
Land required for 3785 m³/d	65 ha	8 ha	24 ha
Land cost, $/ha	$9900	$9900	$9900
Net crop revenue, $/ha	$260	No crop	0

TABLE 9.17.   Cost comparisons of land treatment systems, cents/m^3.[24]

	Flow = 1893 m^3/d			Flow = 189 250 m^3/d		
	Capital	O&M	Total	Capital	O&M	Total
Slow rate	32	8	40	12.8	2.4	15.2
Rapid infiltration	24	6	30	6.7	1.8	8.5
Overland flow	30	8	38	9.7	2.7	12.4

SUMMARY OF TREATMENT PROCESSES

Land treatment processes can be designed to be efficient, cost-effective alternatives for nutrient removal. They can individually, or in combination remove nitrogen, BOD, phosphorus, and trace nutrients from wastewater.

Slow rate land treatment is very efficient in removal of nitrogen, BOD, phosphorus, and trace nutrients. The design basis for the removal of nutrients is well understood.

Design of rapid infiltration for nitrification and BOD removal is well established. Design for total nitrogen removal and long-term phosphorus removal is less common and design procedures are still being developed.

Overland flow can be designed for both BOD and nitrogen removal. Phosphorus removal is limited unless chemical additions or special designs are provided.

REFERENCES

1. Sanks, R. L., et al., "Engineering Investigation for Land Treatment and Disposal." In "Land Treatment and Disposal of Municipal and Industrial Wastewater." R. L. Sanks and T. Asano (Eds.), Ann Arbor Science, Ann Arbor, Mich. (1976).
2. McCarty, P. L., "Biological Processes for Nitrogen Removal: Theory and Application." Proc. Twelfth Sanit. Eng. Conf., Univ. of Ill., Urbana (1970).
3. "Process Design Manual for Land Treatment of Municipal Wastewater." U.S. EPA Technology Transfer. EPA 625/1-77-088 (Oct. 1977).
4. Firestone, M. K., "Biological Denitrification." In "Nitrogen in Agricultural Soils." F. J. Stevenson (Ed.), Monograph No. 22, Am. Soc. of Agronomy, Madison, Wis. (1982).
5. Overman, A. R., "Wastewater Irrigation at Tallahassee, Florida." U.S. EPA, EPA-600/2-79-151 (Aug. 1979).
6. Jenkins, T. F., and Palazzo, A. J., "Wastewater Treatment by a Prototype Slow Rate Land Treatment System." U.S. Army Corps of Engineers, CRREL Report 81-14 (Aug. 1981).
7. McKim, H. L., et al., "Wastewater Applications in Forest Ecosystems." U.S. Army Corps of Engineers, CRREL Report 82-19 (Aug. 1982).
8. "Process Design Manual for Land Treatment of Municipal Wastewater." U.S. EPA, EPA-625/9-81-006 (Oct. 1981).
9. Jewell, W. J., et al., "Limitations of Land Treatment of Wastes in the Vegetable Processing Industries." Cornell Univ. Ithaca, N.Y. (1978).
10. Crites, R. W., "Land Treatment and Reuse of Food Processing Waste." Proc. of the Ind. Wastes Symposia. Water Pollut. Control Fed. Conference, St. Louis, Mo. (Oct. 1982).
11. Smith, J. H., et al., "Treatment of Potato Processing Wastewater on Agricultural Land: Water and Organic Loading, and the Fate of Applied Plant Nutrients." In "Land as a Waste Management Alternative." R. C. Loehr (Ed.), Ann Arbor Science, Ann Arbor, Mich. (1977).
12. "Pollution Abatement in the Fruit and Vegetable Industry." U.S. EPA, Environ. Cen. EPA-625/3-77-007 (July 1977).
13. Perry, L. E., et al., "Pilot-Scale Overland Flow Treatment of High Strength Snack Food Processing Wastewaters." Proc. of the Nat. Conf. on Environ. Eng. ASCE, Atlanta, Ga. (July 1981).
14. Carlson, R. R., et al., "Rapid Infiltration Treatment of Primary and Secondary Effluents." J. Water Pollut. Control Fed., **54,** 270 (1982).
15. Metcalf & Eddy, Inc., "Wastewater Engineering: Treatment, Disposal, Reuse." Second Ed., McGraw-Hill Book Co., New York, N.Y. (1979).
16. Stumm, W., and Morgan, J. J., "Aquatic Chemistry." Wiley Interscience, New York, N.Y. (1970).
17. Spyridakis, D. E., and Welch, E. B., "Treatment Processes and Environmental Impacts of Waste Effluent Disposal on Land." In "Land Treat-

ment and Disposal of Municipal and Industrial Wastewater." R. L. Sanks and T. Asano (Eds.), Ann Arbor Science, Ann Arbor, Mich. (1976).

18. Thomas, R. E., et al., "Overland Flow Treatment of Raw Wastewater with Enhanced Phosphorus Removal." U.S. EPA, EPA-660/2-76-131 (1976).

19. Peters, R. E., et al., "Field Investigations of Overland Flow Treatment of Municipal Lagoon Effluent." U.S. Army Engineer Waterways Experiment Station, Vicksburg, Miss. (Sept. 1981).

20. Uiga, A., and Crites, R. W., "Relative Health Risks of Activated Sludge Treatment and Slow Rate Land Treatment." J. Water Pollut. Control Fed., 52, 2865 (1980).

21. Baillod, C. R., et al., "Preliminary Evaluation of 88 Years of Rapid Infiltration of Raw Municipal Sewage at Calumet, Michigan." In

"Land as a Waste Management Alternative." R. C. Loehr (Ed.), Ann Arbor Science, Ann Arbor, Mich. (1977).

22. Pound, C. E., et al., "Long-Term Effects of Land Application of Domestic Wastewater—Hollister, California, Rapid Infiltration Site." U.S. EPA, EPA-600/2-78-084 (April 1978).

23. Leach, L. E., et al., "Summary of Long-Term Rapid Infiltration System Studies." U.S. EPA, EPA-600/2-80-165 (July 1980).

24. Crites, R. W., et al., "Land Treatment Versus AWT—How Do Costs Compare?" Water and Wastes Eng., 16, 16 and 16, 51 (1979).

25. Crites, R. W., "Slow Rate Land Treatment, A Recycle Technology." U.S. EPA, Office of Water Program Operations (1982).

26. Reed, S. C., et al., "Cost of Land Treatment Systems." U.S. EPA, MCD-10, EPA-430/9-75-003 (Sept. 1979).

Index